358	Williams, James N. (James Napier), 1929-.
.415	The Plan / by James N. Williams. -- Stittsville, Ont.
0971	: Canada's Wings, c1984.
Wil	222 p. : ill. ; 29 cm.

Includes index.
Subtitle on cover: Memories of the British
Commonwealth Air Training Plan.
02336596   ISBN: 0920002293 :

1. British Commonwealth Air Training Plan - History.
I. Title.

Student Pilot, RCAF:
I remember our Elementary School and the long rows of yellow Tiger Moths. And when we walked out with our instructors there was always one of the mechanics standing there to help us get our straps and then to get us started.

THE PLAN
by James N. Williams

CANADA'S WINGS, INC.

Copyright © 1984 by James N. Williams

No part of this book may be reproduced or transmitted in any form by any means, electric or mechanical, in print or electronic media, including photocopying and recording, or by an information storage or retrieval system, without the express written consent of the author, except for brief passages quoted by a reviewer.

Canadian Cataloguing in Publication Data

Williams, James N. (James Napier), 1929-
The Plan : memories of the British Commonwealth Air Training Plan

Includes index.
ISBN 0-920002-29-3

1. British Commonwealth Air Training Plan—History. I. Title.

D768.15.W54 1984   358.4'15'0971   C84-090097-X

Printed and bound in Canada by John Deyell Company

Canada's Wings, Inc.
Box 393, Stittsville
Ontario, K0A 3G0
Canada

Note to Readers

This book is not intended to be a historical document, but rather a collection of the thoughts, impressions, and memories of the men and women who were in some way involved with the BCATP. Because the memory is an imperfect instrument, there may be an occasional error of fact in the stories related in this book. It must also be noted that the stories used as photograph captions do not necessarily describe the person, scene or event in the corresponding picture.

CHAPTERS

1 The Beginnings   18
2 The Fledglings   50
3 The Pros   88
4 The Specialists   112
5 The Visitors   138
6 The Friends and The Faithful   164
7 The Rewards and The Realities   188

PHOTO FEATURES

The Challenge   8
The Ground Staff   40
The Airmen   78
The Civilians   102
The Winter   128
The Good Times   154
The Crunches and The Crashes   178

Introduction   4
Epilogue   219
Index   220
Acknowledgments   222

INTRODUCTION

On the second day of March 1939, at the airport at Vancouver, British Columbia, Sgt R.C. Davis of the Royal Canadian Air Force's No. 1 (Fighter) Squadron opened the throttle of the newly assembled Hawker Hurricane. The aircraft roared, veered off the runway, sliced through the starboard wing of a Ford Trimotor, and destroyed itself in a ball of flame. Somehow pilot Davis escaped.

Squadron Leader Elmer Fullerton, Commanding Officer of the unit, had been assigned the task of converting the RCAF's handful of fighter pilots to their new Hurricane equipment. Fullerton himself had flown a sister Hurricane without incident just a few days before, but with Sergeant Davis's crash, the shortcomings in pilot training for this latest venture in Canadian military aviation became graphically evident. Davis had experience in flying Armstrong Whitworth Siskin fighters, the eminently obsolete biplane in use in the RCAF for almost a decade, but now, with the threat of war in Europe, the Canadian government had ordered 24 Hawker Hurricanes, the most modern front-line fighter in the Royal Air Force. It was painfully obvious that the Hurricane was vastly different from the Siskin.

S/L Fullerton, with more than two decades of military and civilian flying experience to his credit, had found little difficulty in handling the new fighter, but he had pioneered before: in 1921 Elmer Fullerton had joined the Imperial Oil expedition to Fort Simpson, North West Territories. Equipped with two Junkers F-13 aircraft on skis, Fullerton and his companions performed the first winter flying ever attempted in Canada's arctic regions. The group encountered deep snow, rough ice, damaged skis, an engine overhaul, and broken propellers. Fullerton returned from that expedition flying a Junkers with the now legendary hand-carved prop made from sleigh boards and moose glue. Hence, although the pioneering Fullerton could adapt to the Hurricane without difficulty, very few pilots of the RCAF in the spring of 1939 had even come close to a modern front-line fighter. Fewer still had ever flown anything that even faintly resembled one.

So how was it that many years later, Jack Mahr, who flew Hurricanes during the Second World War, could recall something like this? "A fine aircraft; they really were pussy-cats. Manoeuvrable, docile, lots of feel on the elevators, easier to fly than a Harvard. We always used to say, 'If you can fly a Harvard you can fly anything'." The answer: along with 137,738 other airmen from many different countries, Carr had spent almost two years of his life progressing through the schools of The British Commonwealth Air Training Plan.

Ground Staff, RCAF: *Young Davis was lucky to save his skin. Hurricane 312 burned up clean after it had cut right through the wing of the Yukon Southern Trimotor CF-BEP. That's the ship that the Air Force had used in 1931 as the support unit for the Siskins. '31 was a big summer of air pageants, and they went right across the country demonstrating to huge crowds everywhere they stopped.*

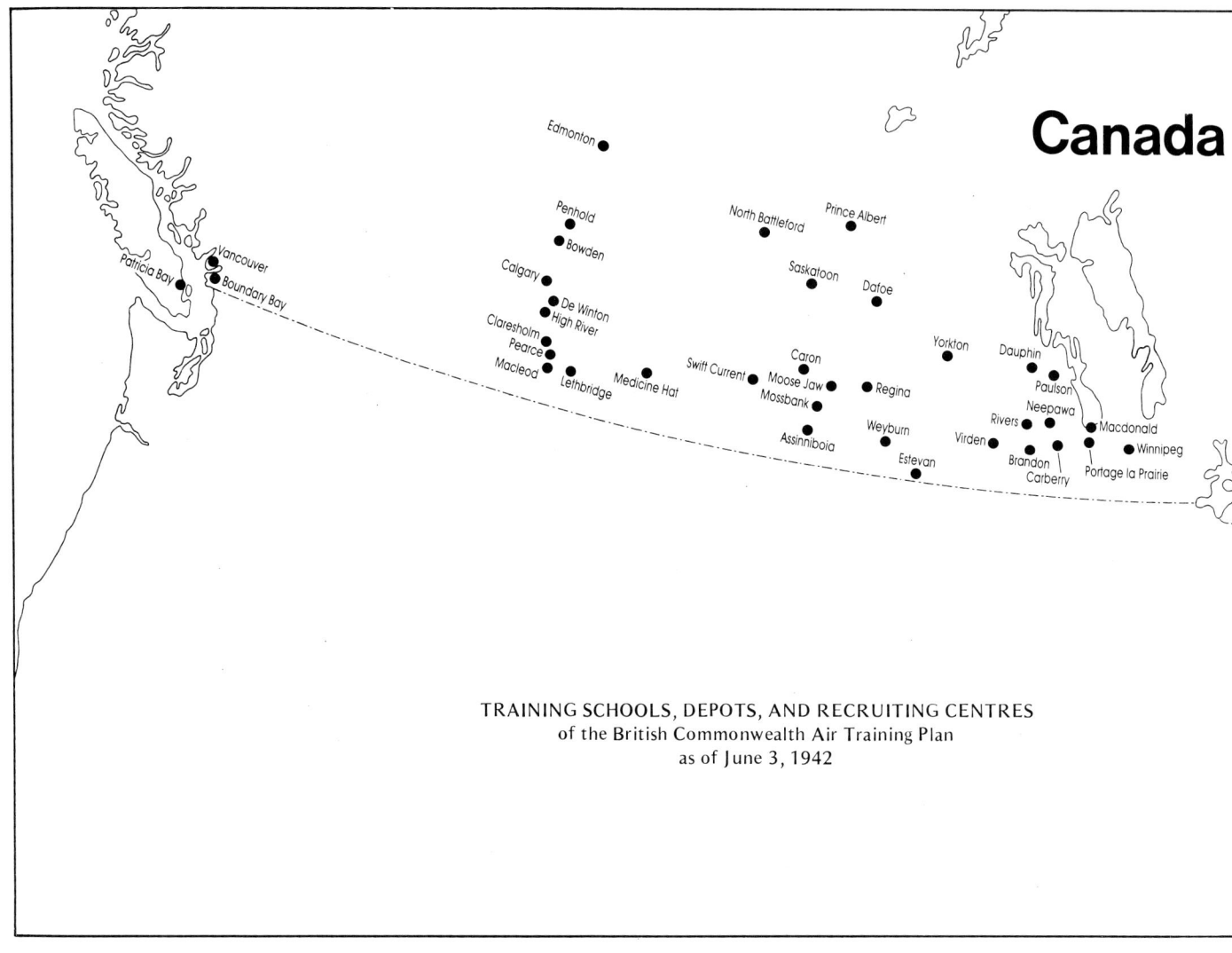

TRAINING SCHOOLS, DEPOTS, AND RECRUITING CENTRES
of the British Commonwealth Air Training Plan
as of June 3, 1942

No 4 TRAINING COMMAND, Headquarters Calgary

Recruiting Centres: No 5 Regina; No 3 Edmonton; No 2 Calgary; No 1 Vancouver. *Manning Depot:* No 3 Edmonton. *Initial Training Schools:* No 2 Regina; No 4 Edmonton. *Elementary Flying Training Schools:* No 5 High River; No 15 Regina; No 16 Edmonton; No 18 Boundary Bay; No 31 De Winton; No 32 Bowden; No 33 Caron; No 34 Assiniboia; No 36 Pearce. *Service Flying Training Schools:* No 3 Calgary (Currie Barracks); No 7 Macleod; No 15 Claresholm; No 32 Moose Jaw; No 34 Medicine Hat; No 36 Penhold; No 37 Calgary (Municipal); No 39 Swift Current; (Admin by No 2 TC) No 41 Weyburn. *Air Observers Schools:* No 2 Edmonton; No 3 Regina. *Bombing and Gunnery Schools:* No 2 Mossbank; No 8 Lethbridge. *Technical Detachment:* No 16 Edmonton. *Equipment Depot:* No 11 Calgary. *Wireless School:* No 2 Calgary. *Repair Depot:* No 10 Calgary (Currie Barracks). *X Depot:* No 14 Regina. *Link Trainer Instruction School:* No 2 Regina. *Operational Training Unit (WAC):* No 32 Patricia Bay.

No 2 TRAINING COMMAND, Headquarters Winnipeg

Recruiting Centres: No 4 Saskatoon; No 6 Winnipeg. *Manning Depot:* No 2 Brandon. *Initial Training School:* No 7 Saskatoon. *Elementary Flying Training Schools:* No 2 Fort William; No 6 Prince Albert; No 14 Portage la Prairie; No 19 Virden; No 35 Neepawa. *Service Flying Training Schools:* No 4 Saskatoon; No 10 Dauphin; No 11 Yorkton; No 12 Brandon; No 33 Carberry; (Admin by No 4 TC) No 35 North Battleford; No 38 Estevan. *Air Navigation School:* No 1 Rivers. *Air Observers Schools:* No 5 Winnipeg; No 6 Prince Albert; No 7 Portage la Prairie. *Bombing and Gunnery Schools:* No 3 Macdonald; No 5 Dafoe; No 7 Paulson. *Technical Detachment:* No 15 Winnipeg. *Equipment Depot:* No 7 Winnipeg. *Repair Depot:* No 8 Winnipeg. *Wireless School:* No 3 Winnipeg.

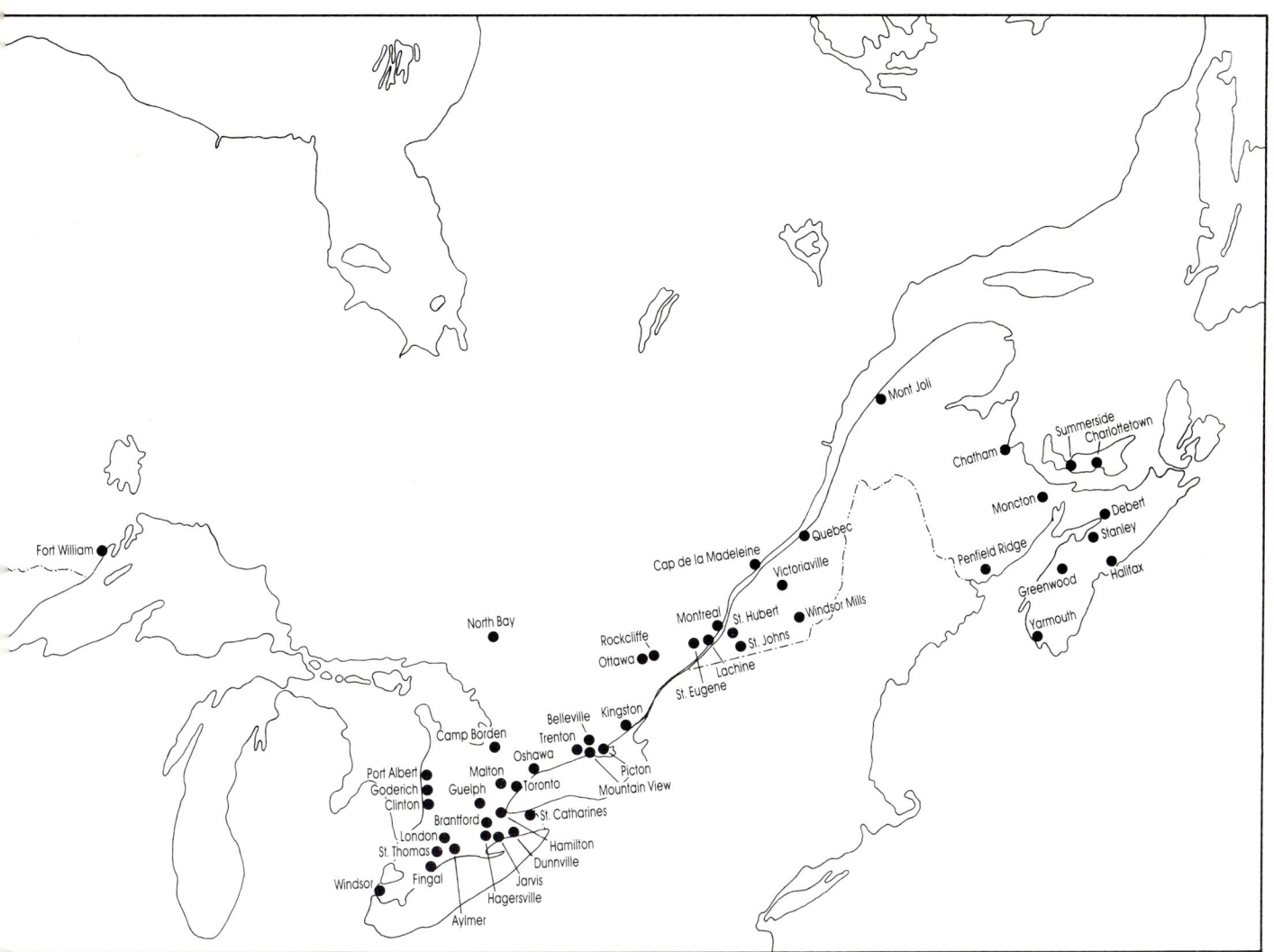

No 1 TRAINING COMMAND, Headquarters, Toronto

Recruiting Centres: No 7 North Bay; No 8 Windsor; No 9 London; No 10 Hamilton; No 11 Toronto. *Manning Depot:* No 1 Toronto. *Women's Division:* No 6 Toronto. *Initial Training Schools:* Nos 1 & 6 Toronto; No 5 Belleville. *Central Flying School:* Trenton. *Elementary Flying Training Schools:* No 1 Malton, No 3 London; No 7 Windsor; No 9 St Catharines; No 10 Mt Hope (Hamilton); No 12 Goderich; No 20 Oshawa. *Service Flying Training Schools:* No 1 Camp Borden; No 5 Brantford; No 6 Dunnville; No 14 Aylmer; No 16 Hagersville; No 31 Kingston. *Air Navigation Schools:* No 31 Port Albert; No 33 Hamilton. *Air Observers Schools:* No 1 Malton, No 4 London. *Bombing & Gunnery Schools:* No 1 Jarvis; No 4 Fingal; No 6 Mountain View; No 31 Picton. *Technical Training School:* St. Thomas. *Technical Detachment:* No 12 Toronto. *AID Inspectors School:* Toronto. *Equipment and Accounting School:* St. Thomas. *Administration School:* Trenton. *Air Armament School.* Mountain View. *Equipment Depot:* No 1 Toronto. *Repair Depot:* No 6 Trenton. *Link Trainer Instruction School:* No 1 Toronto. *Composite Training School:* Trenton. *X Depot:* No 13 Camp Borden. *Wireless School:* No 4 Guelph. *School of Cookery:* Guelph. *Radio School:* No 31 Clinton. *School of Aviation Medicine:* Toronto.

No 3 TRAINING COMMAND, Headquarters Montreal

Recruiting Centres: No 16 Halifax; No 15 Moncton; No 14 Quebec; No 13 Montreal; No 12 Ottawa. *Manning Depot:* No 4 Quebec; No. 5 Lachine. *Women's Division:* No 7 Rockcliffe. *Initial Training School:* No 3 Victoriaville. *Elementary Flying Training Schools:* No 4 Windsor Mills; No 11 Cap de la Madeleine; No 13 St Eugene; No 17 Stanley; No 21 Chatham; No 22 Quebec. *Service Flying Training Schools:* No 2 Ottawa; No 8 Moncton; No 9 Summerside; No 13 St. Hubert. *Air Observers Schools:* No 9 St Johns, Quebec; No 8 Quebec; No 10 Chatham. *Air Navigation Schools:* (both EAC) No 2 Pennfield Ridge; No 32 Charlottetown. *Bombing & Gunnery School:* No 9 Mont Joli. *Technical Detachment:* (AFHQ) No 14 Ottawa; (EAC) No 17 Halifax. *School of Aeronautical Engineering:* Montreal. *Equipment Depot:* No 12 Montreal. *General Reconnaissance School:* (EAC) No 31 Charlottetown. *Wireless School:* No 1 Montreal. *Equipment Units:* No 17 Ottawa; (EAC) No 18 Moncton. *Operational Training Units:* (All EAC); No 31 Debert; No 34 Yarmouth; No 36 Greenwood. *Movements Group HQ:* (EAC) No 11 Halifax. *Y Depots:* (both EAC) No 1 Halifax; No 2 Moncton. *RAF Personnel Depot:* (EAC) No 31 Moncton. *Repair Depot:* No 9 St Johns, Quebec. *Conversion Training Sqn:* Rockcliffe. *Port Transit Unit:* (EAC) No 1 Halifax.

THE CHALLENGE

Ground Staff, Civilian: *We got there in September and there were no barracks, no dining hall, and th hangars were only half finished. The first two months were tough—training the ground people and gettin things organized—but once the crews were into their stride the only real problem we had was the mu*

Ground Staff, RCAF: *We were pushing everything; the mud just seemed to eat things up.*

Ground Staff, RCAF: *They sent our aircraft in before we were ready for them, and our real problem was to figure out how to get the buggers from the field to the hangar.*

TWENTY TO A ROOM

PR Officer, RCAF: Suddenly they had to clear the bush and build all these schools, and at the same time they had to train more people to train others. There was housing and cooking on massive scales. The Air Force people had never had any requirement to do this; in the First War it had been done largely by the British because all the training was over there.

Ground Staff, RCAF: I was posted to a station that wasn't really open. When I got there I found a sergeant and three airmen from Supply. A shipment of beds had arrived and they were unloading these from a boxcar and setting them up in the barracks.

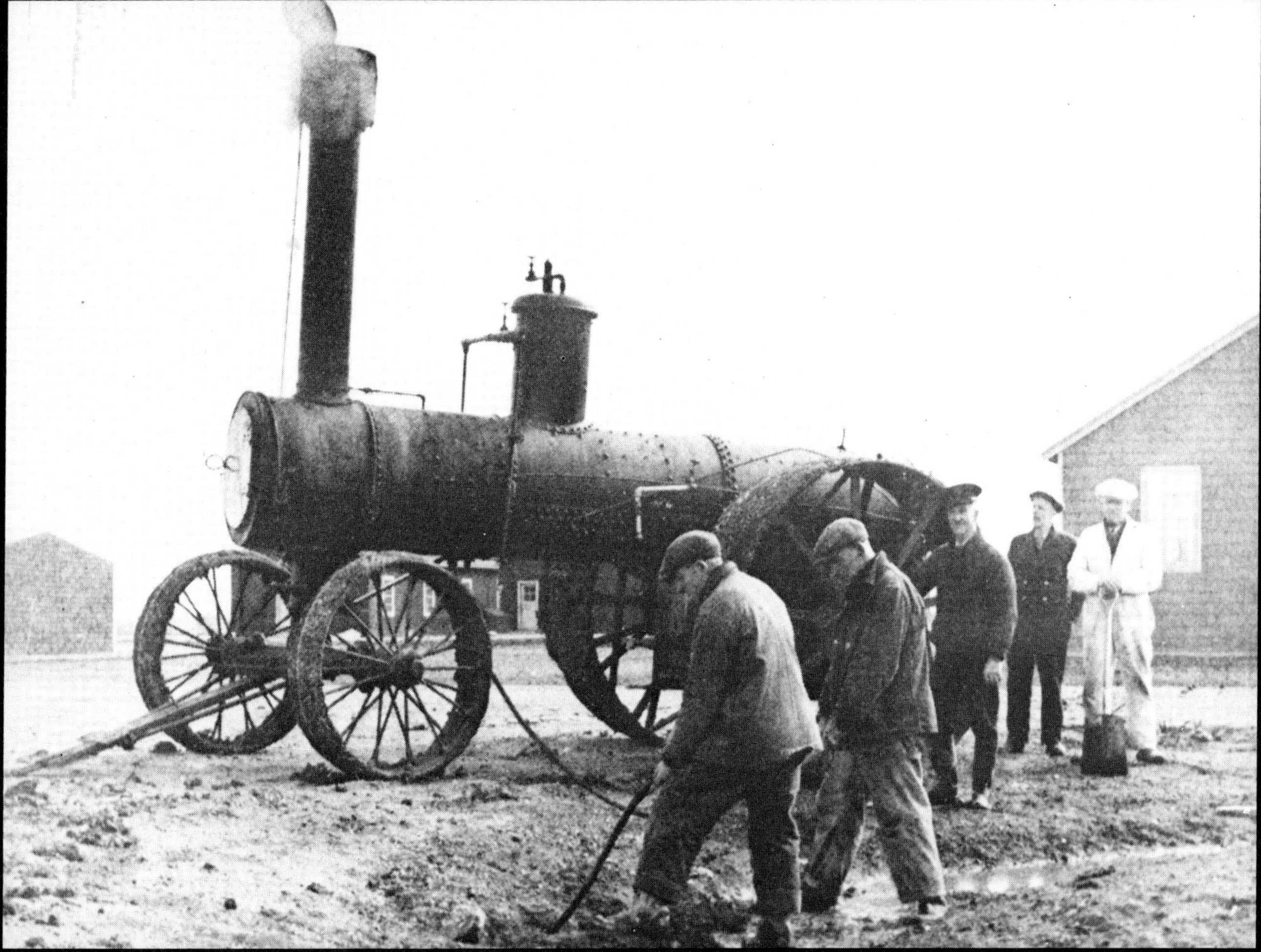

Ground Staff, RCAF: *I don't know where in hell they found the equipment; some of it sure looked like it'd been around for a while.*

WITH A COAL STOVE

Ground Staff, Civilian: The students didn't come in until we had all the aircraft there and everything was set up. Then they had ground school courses for so many weeks before they got on to the actual flying.

Student Pilot, RCAF: When our Elementary School started we were twenty to a room with a coal stove in the middle and we had to feed this bloody thing all night. It was no hotel! One of the flying clubs had formed into a company to operate the EFTS, and the whole thing was run by civilians. Even the instructors were civilians.

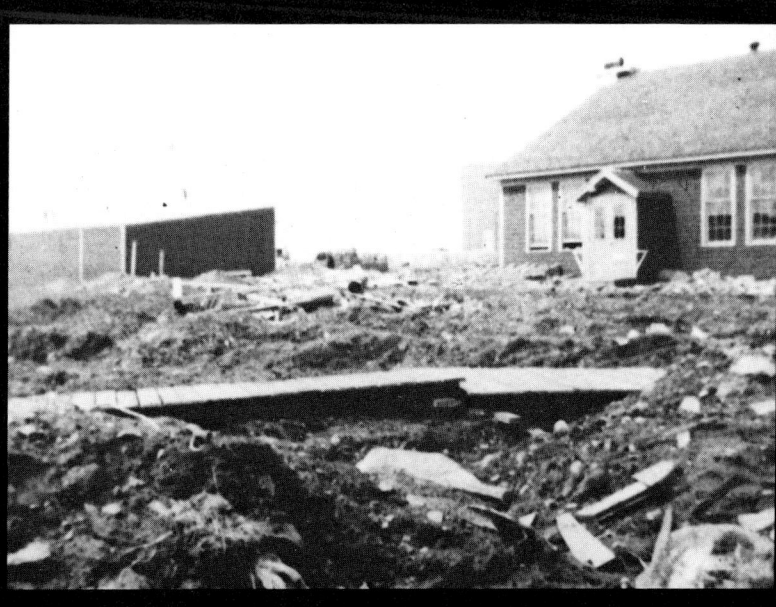

Warrant Officer, RCAF: *Our station was an ape hole. It was all mud and all the ditches were open for burying the water and sewer lines. The buildings were not completed; it was a real mess.*

Building Contractor, Civilian: *We had worked all winter along with the Department of Transport, and considering the weather, we'd made pretty good progress. But when spring arrived we sure weren't prepared for either aircraft or students. The hangars were built, but it was a sea of mud—gumbo, the worst clay in the world, that sticks to your feet so you walk like you were wearing a pair of snowshoes.*

Student Airman, RCAF: *I was sent off to Manitoba. The station was just being built and they wanted a number of bodies there to keep it warm and see if things worked. That was just a dreadful experience—the lowest temperatures I've ever seen, and the food ran out. There was nothing but stewed tomatoes for the whole time I was there.*

At the start no buildings were up, so the company put up construction tents and we stayed in those. It was the middle of September when I left, and it was getting pretty cool. Then they had a company that supplied the food and they had a sort of a big dining tent—almost equivalent to a mess hall—and everybody used to come in there to eat

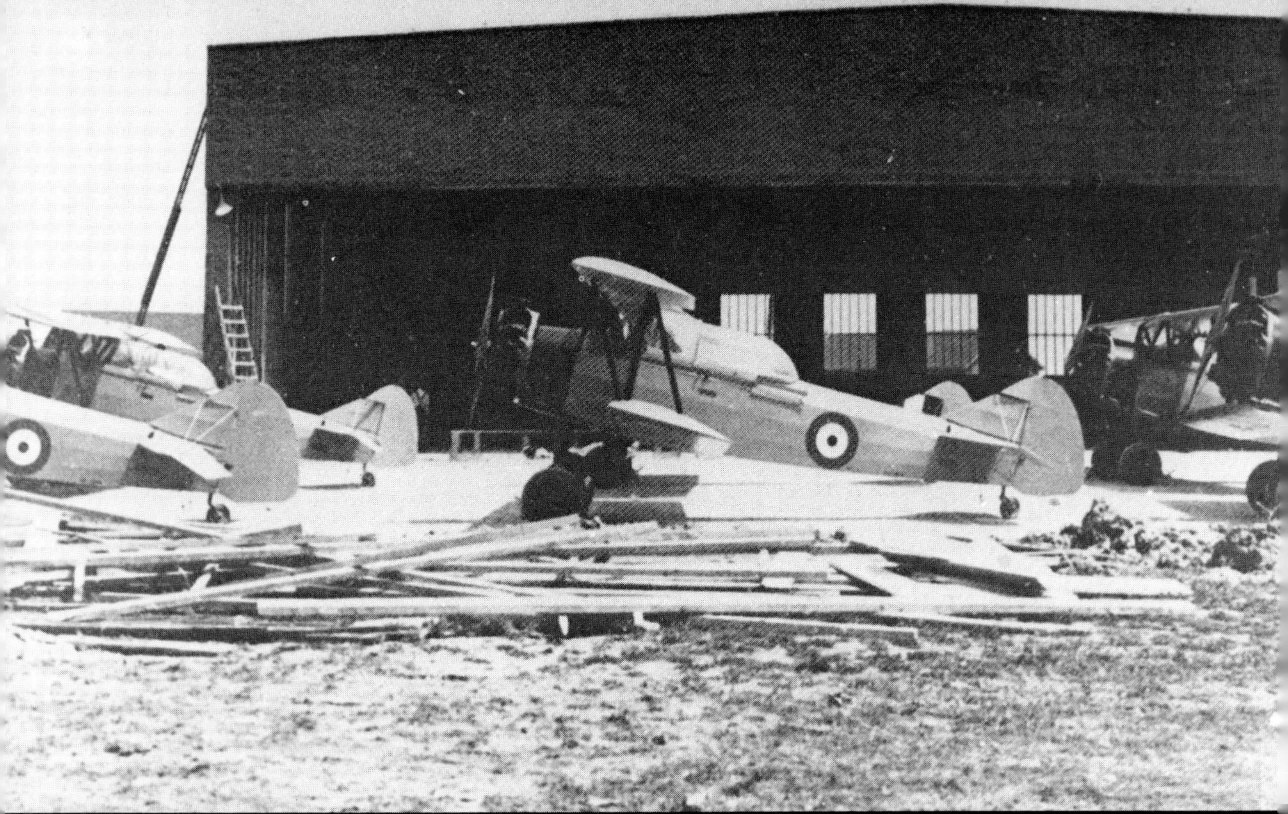

Junior Officer, RCAF: *The important thing was to get a place you could fly off and accommodation, but roads and paths were secondary, as long as you could get there and get the aircraft in the air and down again, that was number one.*

Station CO, RCAF: *It was a fully established station when I got there, a hell of a big layout with a fully equipped ground school, all run by civilians. Gees, there must have been a thousand people there, and the only non-civilians were myself and my Second-in-Command, the Adjutant, and the Medical Office*

Student Pilot, RCAF: *It was the beginning of the thrill of becoming a pilot. You were airborne; you were flying! The first thrill was when the instructor said, "OK, you have control," and then he would talk you through your exercises. Straight and level . . . and a climbing turn . . . and a gliding turn. And one of the big thrills—the first time you land an aircraft entirely on your own.*

Ground Staff, RCAF: *I went to Uplands in Ottawa before the station was opened. The buildings were brand new, and the first week I was there we had our Harvards arrive. Uplands was built on a sand hill and it was a pretty dismal-looking place: the road ended right there; there was nothing beyond; and the only traffic was the Trans Canada Air Lines Lockheed that made the trip from Montreal to Ottawa and then to Toronto two or three times a day.*

OPPOSITE
Ground Staff, RCAF: *Our Ansons came in by train, and we had to assemble them. The engines were crated, but not the airframes—that's just the way they were shipped. It looked like they'd been through the Battle of Britain. Half of them were painted black on the sides and bottom; they were in terrible condition.*

1

The peacetime chores of the twenties
Government promotion of the flying clubs
Bennett attempts to abolish the Air Force
A hint of RAF involvement
Canadian control for Mackenzie King
The big brown envelope arrives
Mr. King's birthday agreement
The RAF shows how
Canada's building boom in the mud
The men at the top
No shortage of volunteers
They were all exactly 39 years old

THE BEGINNINGS

During the First World War 19,000 Canadians enrolled in the ranks of the Royal Naval Air Service, the Royal Flying Corps, and their successor, the Royal Air Force. A significant air training scheme had operated in Canada during those years in which both Canadian and United States servicemen became qualified pilots, gunners, and observers for the war in Europe, but, because Canada had no air force and her political leaders failed to appreciate the future of aerial warfare, the Canadian airmen were absorbed into the ranks of the British air services. For this reason Canada's contribution to the struggle for supremacy in the air during the Great War went largely unheralded and the training of airmen in Canada for that war has been almost overlooked. In 1939, Prime Minister Mackenzie King was adamant: this was not going to happen again.

The Royal Canadian Air Force, which succeeded the Canadian Air Board and the Canadian Air Force, was formed on April 1st, 1924, but the depressed economy of the late '20s and early '30s precluded its expansion.

Early pilot, RCAF: They had a Military Training School at Borden in Ontario in 1924 or '25, something like that. They were running a Summer School there for university engineering students; they were training fifteen or twenty men out of there and they were doing a lot of northern flying, anything to give them a job. They were introducing the idea of a northern forest patrol and of moving surveyors around the country by aeroplane.

Stenographer, RCAF: There was a Director of Civil Government Air Operations who was responsible for civilian flying instructors and he was also responsible for the mapping of the north which was being done by the RCAF. In other words, Royal Canadian Air Force personnel worked for Civil Government Air Operations.

To stimulate aviation, the Department of National Defence proposed the organization of flying clubs and offered attractive subsidies. Sixteen clubs began operations in major centres in 1928, and their membership had reached more than five thousand one year later. These clubs, with the financial backing of the federal government, were responsible for many of the country's early airfields.

Flying Club Executive: The government came up with a scheme late in '23 or '24 that something should be done to revive civil aviation and they proposed to set up flying clubs across Canada. So they sought groups of people in major cities who would undertake to organize such a club.

They were overseas men who founded the club in 1927 or '28, all ex-war pilots. I was in the first group of pupils. The airport was a huge pasture land just north of Portage Avenue [in Winnipeg] that we named Stevenson Field, because he was one of the outstanding flyers in Manitoba; he was in the adventuresome bush country flying.

So much interest was shown that, in 1929, with the promise of an annual federal grant, the Canadian Flying Clubs Association was formed to co-ordinate the clubs. In 1931 the Association promoted the first of the trans-Canada Air Pageants. Together with a selected group of RCAF pilots and aircraft, the Association travelled across the country to most of the major centres which had airports, presenting static displays, fly-pasts, and exciting aerial demonstrations with the newly acquired Siskin fighters.

Stenographer, RCAF: In 1930 I got into the Air Force as a stenographer. At that time you had to have Senior Matric and I didn't have that, but I had been to business college. I was trained to go into accounting and I did the complete course of that, and then the principal said, "The times are going to get tough, and the man that can only do one job is gonna be walkin' the street. Take up shorthand and typing." Gee whiz—I took shorthand and typing. The Air Force couldn't get male stenographers, and on the 7th of September 1930 I'm at Camp Borden as the stenographer to the OC of flying training.

Student Pilot, RCAF: During the late '20s when I was at university the RCAF ran summer training courses for potential pilots. It took three summers to get your Wings, and I got mine in September of 1930. I was summoned by the Chief Flying Instructor, and Old Brookie said, "Campbell, we've been watching you for the last two summers and we've decided to offer you a Commission in the Royal Canadian Air Force." The Air Force was partly under the Army in those days, and the Senior Air Officer, as he was then called, invited General McNaughton to present the Wings. I was one of a group of fifteen or so that got Wings from him.

Generally speaking these graduated pilots dispersed. The RCAF was very small and the intake of regular Commissions per annum was two, three or four, and so there wasn't room for them to be taken on as regular people. Two or three of them went to the Royal Air Force, which was taking short service Commissions at that time, and the rest of them made their way in civil life.

That was about the time when the flight of three Siskin fighters went trans-Canada on an air show basis. They were accompanied by the one and only multi-transport aeroplane we had, which was a Ford Trimotor.

Stenographer, RCAF: They had the Siskins and they had the one Ford Trimotor, and Lofty Roberts flew that across the continent and back. It was more or less to carry spares and groundcrew, and they said if they could land a Siskin then he could land the Trimotor. They got big crowds wherever they went, even at Borden. People would come up from Toronto to watch them, and on a Sports Day —a sort of field day—Bo Riddell would drag a wing tip on the ground. Absolutely perfect judgment—it was dry and you'd see the dust—just a stream of dust all the way around a circle.

Despite this brave flourish of activity, it was generally conceded that "there was absolutely no chance of getting into the RCAF." In 1931 the RCAF training unit at Camp Borden graduated 25 pilots, but there was no room for them in the Air Force.

Junior Officer, RCAF: I was a Pilot Officer and I had had my Wings for eighteen months. At Trenton we had a labour camp; there were 850 men that came in and gave their day's work for food and a place to sleep—and twenty cents a day, the equivalent of five cigarettes. One morning the CO summoned me to the front office and said, "Campbell, you're in charge of messing for the unemployment project." Here I was, a kid, wet behind the ears, *very* lucky to have a job, and I went off and looked after the feeding of all these men. We were on Army rations and it was completely inflexible: if you didn't finish your carrots you couldn't turn the balance in for potatoes; if you didn't use all your salt you couldn't get more pepper. You had to take it or leave it—completely inflexible.

Prime Minister's Staff Member, 1939: Of course, Bennett abolished the RCAF when he was Prime Minister in the 30s. There were a great many Canadians who went to Britain and went into the Royal Air Force, and the Canadian government didn't discourage this a bit.

Stenographer, RCAF: We didn't have an Air Force hardly in 1932. They practically wiped it out, and the men went back to civvy street and were put on a list as reserve. Then there was a group of fully trained instructors—possibly a hundred of them and a lot of those with top categorization—that were let out with the cut. A number of those went to flying clubs.

Pilot, RCAF: *All our work was with floatplanes. There weren't too many airports in those days, and besides, if we had engine trouble, there were lots of lakes handy.*

WE WERE ALL UP THROUGH THE NORTH

Aero Engineer, RCAF: In 1928 after I went to High River, we were doing forestry patrol. We were using Gipsy Moths at that time, flying out over the forests, and when a fire was spotted we would fly to the nearest ranger station and drop a bag with a message in it. We also used carrier pigeons; if we were forced down we released two birds with a message on their leg. They would get back to the loft, and our station would then have the location of the down aircraft.

Stenographer, RCAF: During the '30s Rockcliffe near Ottawa was a base for seaplanes. They were quiet in the wintertime, but as soon as the spring came they were all up through the north photographing and mapping the country. There was a field there, but there were no paved runways anywhere else. There was no Uplands then, so the only aerodrome in Ottawa was Rockcliffe. It was an Air Force Station, but it was also used a bit for civilian flying.

Aero engineer, RCAF: The real early types we used were the Curtiss HS 2L and the DH Moth. They used to use the HS 2Ls for convoy patrol work out of Halifax in the First War. The US Navy had a bunch of them.

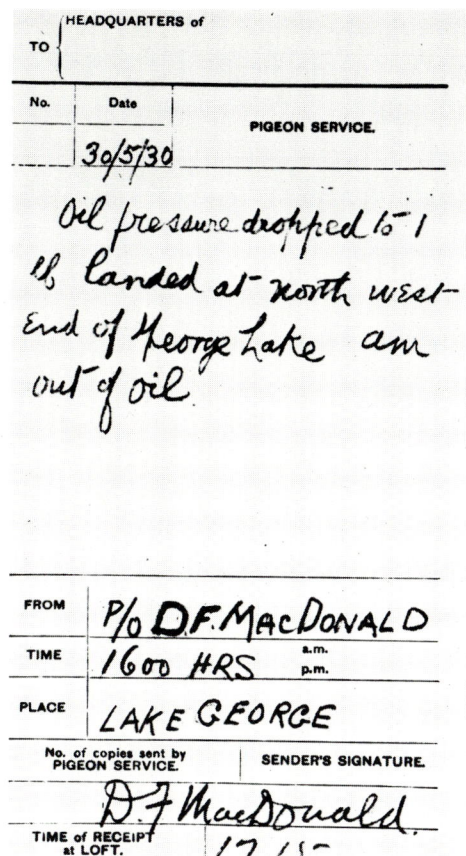

Pilot, RCAF: *Most of our early mapping and patrol work was done with DH Moths. As time went on the planes got more sophisticated, but we never did get any proper radio equipment. The pigeons always seemed a bit of a bother, but it always felt good to know they were with us.*

Officer, RCAF: The RCAF continued to contract until after Trenton was opened, which would be about 1933 or '34. All the non-regular officers were let out, and all that were left were a few people that held regular Commissions, of which I was one—I was very lucky. It happened just overnight during the Bennett regime.

Many of the graduate pilots made their own way to Britain, where, if their qualifications were considered to be of a suitably high standard, they were taken into the RAF. Only small numbers were accepted, but Britain was able to skim the cream. Although pilot training continued in Canada, by 1935 there was still no major increase in RCAF pilot strength.

Pilot, RCAF: At Borden we went onto Atlas aircraft to carry out our sort of operational flying, which was primarily air firing and sand bagging. The Atlas was a single-engine Army co-operation aeroplane which was crewed by a pilot and an air gunner cum wireless operator. It was a biplane, and sand bagging was by guess and by God; over the back, the air gunner would throw it out. That's how "ancient and honorable" we were in those days!

We used to do night flying with the Atlas, and what we used for landing lights were phosphorus flares on the ends of the wings. When you got in the right position of approach you'd flash a button, and on these phosphorus flares would go, under the canvas wings. So when you got down you couldn't afford to stop; you had to keep going to keep the flame from the flares away from the dope and canvas wing tips, or you'd lose an aeroplane!

Canada and the United Kingdom began discussing the future of air training in Canada in 1935. In 1937 Britain asked for an agreement allowing Canada to train up to 125 Canadians annually as pilots for the RAF.

Mackenzie King was wary. Although he liked the idea of training more Canadians as pilots for the RAF, he had not forgotten how the Canadian identity had been lost in the First World War through the absorption of Canada's trained airmen by the British forces. Consequently, King set the British request aside.

Officer, RCAF: I was in Vancouver from '36 to '39 as the Adjutant of an Auxiliary Squadron. That consisted of being a one-man show: you were the flying instructor on Saturdays and Sundays; you had Mondays off; and on Tuesday and Thursday nights you gave lectures on Air Force subjects to Auxiliary

Flying Club Executive: *Around 1923 the government said, "If a group would like to start a flying club, if they will buy one of these De Havilland Moths, then we'll give you a second one. But you must agree to set up a field and hire an instructor, and under proper guidance from the government you will offer an approved training plan."*

officers. In addition, you had the responsibilities of Equipment Officer and Engineering Officer.

Mind you, the operation wasn't very big, and in those days you got a good cross-section type of experience. I was one of three new officers to join in 1931, and the Air Force didn't take any more regular Commissions for another two or three years. My number in the RCAF was 132; I was the one hundred and thirty second officer to join the Air Force, and up until '39 I knew over a hundred of them and dealt personally with the majority of them. We weren't very bloody big!

Some expansion of the RCAF training programs did take place. In January 1938, under what was known as the Trained in Canada Scheme, fifteen candidates for the RAF started their training at Trenton. This could be considered the forerunner of the vastly enlarged plan that was to be implemented in less than two years.

Official, Dept. of Finance: The British were sending a few pupils over to be trained in the RCAF, but it was on a small scale and they were part of the regular RCAF operations. There was room for a much larger operation and there was a role for it, but it just wasn't possible to get agreement between the two governments on anything big enough to be a major force to contribute to what might happen in a war.

Prime Minister King still continued to oppose Canadian training of Canadian pilots in Canada for the RAF, but he did suggest that: ". . . we would agree to co-operate to the extent of all the training space they wish." King then stated that control of such training by the RAF would not be acceptable: "We are prepared to have our own establishments here and to give in those establishments . . . facilities to the British pilots to come and train here. But they must train in establishments which are under the control of the Government of Canada . . . and for which the Ministry of National Defence will be able to answer in Parliament with respect to everything concerning them." *At this time the only two RCAF stations capable of training pilots were at Borden and Trenton, and these two combined were barely capable of meeting RCAF requirements. Nonetheless, within a week of asserting Canada's control over future pilot training in Canada for the RAF, King offered to train "not in excess of fifty per year" to help with RAF requirements. Even this number would require an increase of both staff and aircraft at the two training stations.*

Building contractor: About 1937 or '38 mysterious British money was willing to build big hangars in Winnipeg. Everywhere things were going up, and British pilot trainees began to appear. Residents took them in; we had some staying with us, but we couldn't figure out what in blazes they were going to do with them. They didn't know why they were here either, but they were young fellows who had been offered flying training, and for an unpublished reason the British government was promoting the expansion of the Royal Air Force.

I was prominent as a supplier of building materials, and we just couldn't understand what was going on. But the British knew that a war was coming and preparations had to be made for training—and it was time they got to it.

Recruit, RCAF: I left Wolseley, Saskatchewan in May of 1937. I can remember getting the big brown envelope way out in the sticks—that was a real glorious day. I'd heard about the Air Force in about 1935, that it actually existed in Canada. As a kid on a farm I just stumbled on the fact that this country *did* have an Air Force. I sent them a letter and a few months later they replied saying that the economy of the country was such that they weren't recruiting, period, and they more or less said, "Forget about it; we'll be in touch with you, but don't bother us." With that in mind I took a two-year course in auto mechanics in Regina and then early in the spring of '37 this letter came saying, "OK, we're considering you as an applicant if you can meet the qualifications medically." So a couple of weeks later I was off to Camp Borden.

It was a pretty grim situation for most of the West in those years; in '36 and '37 we were just starting to come out of the big depression. When I left Regina I didn't have the money for even a decent suit of clothes or shoes, and the local merchant grubstaked me on that. He said, "OK, Frank, when you get down to Ontario and the big money's rollin' in, send it to me." I thought I was the only guy from out West that was on his way to Borden, but there were about 130 of us that eventually showed up.

Trainee, RCAF: We were set up at the shops along the hangar line. There were shops for mechanics and airframe people, and machine-shops, and also there were people who were training in administration. Oh, it was a going concern; there were over 100 of us in that first class.

Stenographer, RCAF: That first class of No. 2 TTS in '37 was at least a third of all our technical men when they graduated. That was the start of the build-up; we built up very fast from there. In '38 there was another intake and in the spring of '39 we started taking people in from civvy street. They took hundreds in that were in jobs that sort of related—we took in chaps who were working in garages on cars that never had attended a technical training school. They went right to work in on-the-job training and qualified. In '38 there was a staff of around twenty at the school—that's all—and there were only four officers on the station.

Trainee, RCAF: I was learning to become a fitter. My group had about eight or ten guys and there'd be eight or ten in the next group, and each group had instructors. But you weren't always in the classroom; you still had to be an "Airman-Soldier" you might say, and the rifle drill was important, too. Nineteen or twenty of us were selected from the total of the people there under training for a new classification that was being formed called Instrument Mechanics.

Stenographer, RCAF: Of course we had a pick—no problem finding men. There were all kinds of men wanting to come in and we had our choice, but even at the outbreak of war the Royal Canadian Air Force, including the Auxiliary Squadrons, had a total strength of only about 4000.

Student Pilot, RCAF: *I didn't participate in the operational activities of the Siskin Squadron, but I certainly used to stand back and look at them with respect as to the job they were doing and the manner in which they were developing a sort of understanding on the part of the taxpayers as to what the Air Force was all about.*

Prime Minister's Staff Member: In 1938 the British made a proposal that they should train their own Royal Air Force pilots here in Canada, and we said that we would agree to this only if the stations were Canadian, and commanded by Canadians, and an agreement was made after the first of July, 1938. We refused to let them establish stations and run them themselves, but we said we would be very glad to have them train here under ultimate Canadian control.

Although this was July of 1938, the British never got around to doing anything about it until the war started. Once the war was started then Chamberlain proposed this Commonwealth (they called it "Empire", but we always called it "Commonwealth") scheme—and we made very bloody sure that there was no question whatever that this would be under Canadian control and the organization manned largely by Canadians.

January 1939 saw the beginning of intense negotiations between Canada and Britain and of changes in policy for Canada's pilot training program. Contracts were awarded to eight of the civilian flying clubs to operate Elementary Flying Training Schools. In April 1939, a training course at Borden qualified instructors to teach RCAF procedures.

Student pilot, RAF: In the spring of '39 I applied to get into the Royal Air Force with a short service Commission with an ultimate view of going to Egypt. It was before war broke out that I joined, but then things started to get pretty sticky. I guess they felt that war was imminent. I was posted to No. 1 Fighter Squadron in Calgary, Currie Barracks, and we trained in Puss Moths and Gypsy Moths under a civilian instructor.

Official, Dept. of Finance: Mackenzie King knew that if a war started and if the British got into it that Canada couldn't keep out. In order to keep peace in Canada he wanted to make as few advance commitments as he possibly could, and hence he didn't want to get into these joint training schemes with Britain.

Air training was a natural for Canada in the event of a war in Europe in which we were going to participate. The idea was first talked about back in '36 or thereabouts, but for one reason or another it was not possible to get any sort of agreement between the British and Canadian governments. However, we started on a Sunday morning, the first of September '39, when word came through that the Germans had invaded Poland, and from then on it was day and night in the Finance Department.

Junior Officer, RCAF: At the outbreak of war the important thing was to build up our flying capabilities *here*, and we needed groundcrew for that. There were chaps wanting to get overseas, but the answer to all the old-timers was "Sorry, boys. We can bring somebody off civvy street and have them flying just as quick as you, but we can't bring anybody off civvy street to take your place. You can't go over because the RAF wants us to send aircrew." And so we were then going full out getting groundcrew trained.

Official, Dept. of Finance: At that time it was necessary to fix on programs for the Air Force, the Army, and the Navy, so that we could take them into account in the budget. There was a problem in deciding how much we could do in the first twelve months of the war, and this was settled in a rough sort of way in the first ten days of the war. But only after it had been announced to the Canadian people did the British come through from Prime Minister to Prime Minister with a major proposal for setting up a large Commonwealth training plan for airmen in Canada.

Britain foresaw an annual requirement of 50,000, and because of Canada's geographic location and its easy access to the United States for supplies, it was desirable that the training be carried out in Canada. Also, Canada was not susceptible to harassment by German air activity. The proposal and its apparent magnitude duly impressed Mackenzie King: "The most effective contribution Canada could make is through a great co-operative project to train pilots and aircrew for the Commonwealth Air Forces."

Official, Dept. of Finance: The British were proposing a very big operation on a scale that was quite beyond anything we had ever done or planned before. Then the British Mission got out here and we had our first high level meeting before the end of September.

Prime Minister's Staff Member: King was bitterly disappointed that Chamberlain didn't make his proposal until after the Canadian military program had been set and announced to the Canadian people earlier in the month.

In the years between the wars, more than 400 Canadians joined the RAF, many from the ranks of those trained in Canada. During that period there were more pilots from Canada in the RAF than there were in total in the RCAF. In August of 1939, there were 235 pilots in the whole of the RCAF.

Up to the end of '36 civilian pilots came to Borden and got a flying instructors' category, and then a year or so later he would come back to Borden and get his category raised, until he got the top rating. They came from all over Canada, and they eventually became the nucleus of the instructors for the Elementary Schools of the BCATP. In '31 there was one of the guys that was in there that failed. I remember when Brookes brought him in—George Eric Brookes—he was up to test him out. He came in too low to suit Brookes, and Brookes said, "You know, we were coming in over those darned pines, and I reached down and caught my balls and held them up so they wouldn't hit on the tree tops!"

Prime Minister's Staff Member: The British sent out their delegation to negotiate the deal and to work out all the details, and that lasted from early October to mid-December, which was a hell of a long time. The British at this point . . . if you could say they were taking the war seriously, they certainly weren't taking it very *urgently*, and once they found out they weren't being bombed they got pretty relaxed about the whole thing.

By mid-October, Britain's aircrew estimates had been set at 30,000 a year. This would require some 5000 training aircraft and an increase in RCAF personnel to almost 55,000. It was pointed out that "this was so far ahead of anything we had thought of that everyone who heard the details was quite taken aback by its magnitude." The question was, "How much is it going to cost?"

Official, Dept. of Finance: It was clear that we couldn't produce the aircraft in Canada that would be necessary to run such a big training operation, so it was recognized from the beginning that the British would have to supply the bulk of the aircraft to get the program started. We also recognized that we'd need a whole lot of aerodromes and that we'd have to build a lot of schools and hangars and what not.

Also, the British couldn't foresee any effective early fighting force other than as a part of the RAF. The Canadians, on the other hand, were guided by Mackenzie King and his External Affairs Department, who were much against any plan which looked as though we were committing ourselves ahead of time to fight as part of the British Forces, and so they resisted any proposals to set up in Canada what looked like a part of a British operation.

But we had negotiations on what this thing was going to involve in terms of numbers of schools, the size of those schools, equipment, and all that, and what they would supply, what we would supply, and how the costs would be met. This went on for some weeks and finally we had a series of critical issues that arose as to the identification of the Canadians that were going to go through the schools and whether they were going to wear Royal Air Force uniforms or Royal Canadian Air Force uniforms.

Prime Minister's Staff Member: The person who was most keen about the Canadian identity was Chubby Power, and once the position was put to him, Mr. King supported the feeling that Canadians should be in RCAF uniforms.

These meetings resolved several basic points. The program would be located in Canada. Canada was to train all her own airmen from their initial training through to their preparation in Britain by the RAF as fighter and bomber crews. Australia and New Zealand, which already had sizable training schemes and whose governments believed that for political reasons training must continue in those countries, would provide initial training and elementary flight training for their own students. These students would then be sent to Canada for further training. Each Dominion was expected to pay proportionate amounts for the training of their own airmen. It was also accepted that Britain was to contribute a significant number of training aircraft along with extra engines, spares, and accessories.

Prime Minister's Staff Member: King got fed up with the long delays. Apart from everything else, the British were really very slow about the thing, and Ralston, who was Minister of Finance at the time, was very anxious to make the British pay as much as possible—and they were anxious to pay as little as possible. The thing dragged on and on and King bullied the head of the British delegation. During the final meeting one of them was at a hockey game and Mr. King *summoned* him: "Get him here at once!" It was around midnight before the thing was finally signed.

Official, Dept of Finance: The RCAF had a first rate accountant, Harry Norman, on the financial side. C.D. Howe in his Transport Dept personally took on the big program for acquiring the aerodromes while the RCAF was organizing as quickly as possible the necessary training personnel and recruiting the first batch of trainees. We believed from the beginning that it was going to be necessary to get some British training officers and instructors; whether we had the necessary groundcrew to get such an operation going and keep it going wasn't clear, but the upshot was that it looked like a feasible proposition if the British could supply the aircraft. Then we had to think of the financing of it.

Official, Dept. of Finance: *My guess is that in the last 48 hours before the agreement was signed, Mr. King probably had his birthday in mind—he was a superstitious old cuss—and the agreement* did *get approval and was signed, but it spilled over an hour or two after midnight. We had reached agreement on the central questions before reaching agreement on the status of the graduates and the relations between the RAF and the RCAF and that sort of thing, but the gist of it was that they couldn't make the announcement until his 55th birthday.*

THE OTTAWA EVENING CITIZEN

CANADIAN TROOPS REACH CAMP IN SOUTH ENGLAND;

Pact Signed for Great $600,000,000 Empire Air Training Scheme

Prime Minister's Staff Member: That was the day that the first Canadian troops landed in Britain, and Mr. King was very anxious to get the thing signed, not just because of his birthday, but so that the whole focus would not just be on the arrival of the troops; it would give some evidence that we were doing something else as well.

When the anticipated costs were totalled, Mackenzie King and his government were astounded. Canada's own estimated expenditures for the first year of the war were discouraging enough, but aircrew training was, "after all, a scheme suggested by the British government and for which the British government must be mainly responsible." All trainees from outside Canada would be attached to the RCAF and receive Canadian rates of pay. In general, training methods and standards would be those of the RAF. An advisory board with representatives from Australia, New Zealand, Canada, and the United Kingdom would meet each month.

King considered this to be an undertaking of supreme importance: "It will establish Canada as one of the greatest air training centres of the World. ... The aim in short, is to achieve by co-operative effort, air forces whose co-ordinated strength will be overwhelming."

On December 17, 1939, with the signing of the Agreement, the British Commonwealth Air Training Plan became a reality.

Getting Under Way

By the end of March 1940, total RCAF personnel strength exceeded 10,000, more than double that of August 1939. Nonetheless, if the BCATP was to reach its expected output of 1,500 trained aircrew every four weeks by the end of two years of operations, the increase in RCAF numbers would have to be far greater.

On April 29, 1940, BCATP operations officially started, and the bulk of the energies and efforts of the RCAF were directed toward achieving the goal that had been set: the BCATP would be one of Canada's primary contributions to the war effort.

As the scope and responsibility of the Plan became clear, the RCAF rightly concluded that it did not have sufficient expertise within its ranks to administer such a massive undertaking and therefore requested help from the RAF. The RAF had had a training operation in England for many years, not on the scale now proposed by the Plan, but of sufficient size to have developed a professional approach to large-scale aircrew training. The British were more than pleased to co-operate. Early in 1940 a group of about 100 experienced administrative personnel from the RAF arrived in Canada.

Senior Officer, RCAF: The then Chief of the Air Staff said they wanted assistance from the Royal Air Force in the way of experienced personnel in executive and administrative jobs, so then they were fed into the RCAF structure at pretty well all levels.

Officer, RCAF: There was quite a crowd came over from England, and they helped start the Technical Training Schools. It wasn't just aircrew training in Canada—it was aero engines, airframes, and other ground services. St. Thomas became the main station for that technical training.

Ground instructor, RCAF: I was posted to St. Thomas as an instructor in January 1940 in what they called Flight Routine, that is, in running the engines and servicing the aircraft and training these new people in what to do. Then the RAF sent over a staff of approximately 100 and they took over the training. It would be about March of 1940 that they started coming in.

Ground instructor, RCAF: The RAF fellows were sent out here to set up the Plan in conjunction with what was then the RCAF Reserve—which wasn't very large—and through their collective knowledge they formed the nucleus of the Ground Instructors. They took me off a course for Wireless

Operator Ground after just a few weeks and said, "You're a Morse instructor," because I knew Morse from before the war. At that time anybody who had the least bit of knowledge about anything was an instructor.

PR Officer, RCAF: The Canadians put it up to the British saying, "We can send you some over now, but we see our task as trying to train a *few* men, who will then train a hell of a lot *more* men," and the contribution then would become significant. The British actually said, "We'll hang on and you can give us a whole lot later," because the more trainees you held back here in Canada, the more people you had to train many *more* people. But in the meantime the cost was paid over there in England.

I went to St. Thomas; there were magnificent buildings there and you could go from one to another through underground tunnels. If there wasn't someone who knew the way you could get lost and it would take hours to get back out. It was originally built as a mental hospital and the windows were barred, but when the war broke out they needed a school to train ground personnel, so they made the insane wait.

We were at school eight hours a day and we got every second weekend off. The instructors were good, although when I was there in the first year of the war a lot of them didn't know much more than we did.

It was an odd set-up, nothing like any other Air Force accommodation. The rooms had double bunks, but the walls only went two-thirds of the way to the ceiling . . . and they were like bays instead of rooms; they weren't actually closed in. Being built for an asylum, there were no moving parts on anything; there were no toilet seats—the toilets were never built to put seats on—so you sat on cold porcelain. The buildings were almost in a circle and the centre was used for a Parade Square. It was an ideal set-up for a training school.

We learned hydraulics—flaps and retract gear—and they had parts of aircraft and old engines. There was a conglomeration of aircraft bits and pieces that everybody could train on. It was strictly technical, and you came out of there as a fledgling aero engineer. Because the place had been built as an asylum, the showers were all controlled by one valve, and we used to get scalded by guys walking in an giving the valve a twist.

Part of the course was held in Toronto at Central Tech. Apparently that was the only training school in the country for watchmaking, and they had ordinary daytime students, but then Air Force people came in at four o'clock and had classes till eleven. They had a jeweler's lathe set up and we shared our work benches with WACs, the Army girls. We were learning to do very small work so we could repair gunsights and cameras.

Prime Minister's Staff Member: *Mackenzie King was very keen about the Air Force, both before the war and at the beginning of the war, and he had an ulterior motive for this. He was scared to death of a big Army because it would probably lead to an agitation for conscription. But he had an utterly unmilitary mind. Nobody could have been a great head of a war government (and King was a great head of a war government) who knew so little about the Armed Forces. He could never even learn the military ranks.*

I only had Grade 11 and I was a little short on math and physics. They had what they called WETP—War Emergency Training Program—which we affectionately called "wet pee school"—and there was a group of us went to that before going anywhere else and we started a three-month course in Saskatoon which was run by civilians. There we were given good basic mathematics and courses in internal combustion engines, ignitions, propellers, and the use of welding equipment, so that when we went to St. Thomas we weren't just off the street.

The responsibility for preparing the actual surfaces of the new training fields fell to the Department of Transport, but the RCAF took on the task of erecting the buildings. Formerly this had been the job of the Army Engineers, but the projects required by the Plan were of such proportions that a special unit of the RCAF was formed to look after building construction. Each school was a self-contained operation, requiring living accommodation and mess halls for students, instructors, and ground staff, administrative space, recreational facilities, repair shops, power and heating plants . . . and hangars.

Engineering Officer RCAF: In the mid-twenties I joined a horse cavalry regiment and continued on there until the war started. Our job then was to guard the railway bridges from Montreal to Halifax over which all the stuff was being passed to head overseas. We were on the job for a couple of months when word was received at Headquarters in Saint John: what was an engineer doing in the New Brunswick Hussars when the Air Force wanted to increase its engineering staff because it was starting up the British Commonwealth Air Training Plan?

I then went to Ottawa and switched over to the Air Force in January of 1940. The Headquarters of the Air Force was then in the Jackson Building on Bank Street and the officer in charge of personnel was Wing Commander Wilf Curtis. He said, "We've got four Engineering Officers here now and we've been able to persuade Dick Collard (who was president of a large construction firm in Winnipeg) to come and head up our new Engineering Group. So if you head over to Cartier Square, you'll find Wing Commander Collard, and he's got two or three chaps around him."

This group of four or five people was the nucleus of what became the Works and Buildings Branch of the Royal Canadian Air Force. They also were able to bring back into the service fifteen or twenty holdovers from the First War who'd had experience with the RAF, and *they* had all kinds of thoughts and ideas which were helpful. As well, we appealed to the Engineering Institute of Canada and the various engineering groups in the provinces, and they in turn passed on the word to their members that we needed help.

Officer, RCAF: They brought in a lot of very experienced people in the engineering field and gave

In the fall of '39 I was on the draft with 110 Squadron as an instrument mechanic. We were closely affiliated with the Army and the work they had to do, and one of the jobs was to observe the barrage balloons flying over London and other major cities. Word came through that we were being sent back to Canada to take part in this British Commonwealth Air Training Plan—whatever that was! I didn't have a clue what it was, but it was explained that we'd be setting up shops in different training schools throughout the country.

them Commissions. Later on we had the Works and Buildings Branch of the Air Force, and that was on an equal footing with Operations and Training.

Assistant, Dept. of Munitions and Supply (M & S): The Air Force would decide where they wanted the airports, and M & S would go out for tender for their construction and award the contract and supervise. In addition to that, there were the buildings that were required, hangars for maintenance and accommodation for the staffs and the men that were being trained.

Officer, RCAF: Every constituency across Canada was shouting for a military base. You know what a couple of thousand people meant to a town—and politicians do their damnedest to get a crowd in—but then they are miserable when the people get there. All they are interested in getting is their money.

Trainee, RCAF: But just imagine what would happen with the influx of a bunch of students and groundcrew to a place like MacLeod—or Yorkton, or Dauphin, or Mossbank. The money that would be spent there would make a tremendous impact on most communities. The other thing was that many outsiders inter-married with girls from that community, and so the Plan had a tremendous effect across the whole country.

Engineering Officer, RCAF: The buildings were done as an ordinary contract job. Big construction firms would start up a group within themselves that could move in and do this. Commonwealth Construction and Marwell Construction did all kinds of our work out West. In the East it was Modern Construction in Moncton, the Foundation Company was a huge concern, and Vickers Shipbuilding had a branch which did work of that kind. In the electrical field there was Comstock.

Those people really pitched in and worked all hours, and the companys would loan us the services of staff. If we were jammed at Headquarters for a particular time I would call up somebody that I knew in a firm and say, "Look, boys, is there a chance of having Harry Smith here for a month or so, because that would be a great help?"

The selection of sites was the responsibility of Department of Transport, but final approval rested with the RCAF. Location partly depended on the proposed use of the school within the Plan, but a prime factor was the field's potential for use at the end of the war. Locations also depended on the nature of the surrounding terrain.

IT ALL HAPPENED IN OTTAWA

Engineering Officer, RCAF: All the original planning occured at HQ in Ottawa. We worked very closely with the Department of Transport; they had teams of people that went all across the country to pick areas that could most speedily be transformed into training stations. We had a standard layout for an EFTS, for an SFTS, for B and G, and one for ANS and AOS. As the war went on there were radar training spots which were manned by RAF people.

The hangars were made almost all of wood. We had a standard 110-foot hangar for the small planes, and a double hangar for the larger planes was two of these with extra supports at the centre to give the wide span. Then we had lean-tos on either side in which were the workshops and offices.

Building Contractor: The hangars were designed in a central area in Ottawa and they came prefabricated from large producers of heavy timber. They were all timber; beautiful timbers in those roofs, lots of fir, four by twelve, four by eighteen, even four by twenty. There was a shortage of steel, they needed that for shells, and so the wood structure fitted in admirably with Canada's production. And that's how those trusses were made—just wood and some bolts.

Construction Helper: The hangars came in by rail and then they were trucked out just as you'd truck a load of lumber. They were all shipped pre-cut and all we did was bolt the skeleton together, nail on the siding, and shingle it. The doors were shipped in ready-built. We could put up a hangar in no time.

Heating Contractor: Those hangars were all steam heated and we had to get that in before the first winter hit. Everything was done from central boilers, but there wasn't much insulation used anywhere in those days, so it was pretty uncomfortable in real cold weather. In some of those barracks we used a thing called an iron fireman. They bought a whole bunch of those and stuck them in a bit of a shack at one end of the building. They didn't work too good because they often filled the whole place with black soot and the boys would get up in the morning with black all over their faces.

PR Officer, RCAF: *Air Force Headquarters had small teams of experts. Many of them were civilians, because they knew how to deal with industry. Every morning I'd pass a couple of these fellows sitting in a little office about the size of a chair, and they didn't seem to be doing much but looking out the window, but they had charts and things out on the table. When I got to know them I asked about this, and they said they were thinking, "What is the next thing we're going to need and how many? And where are we going to get them:"*

Ground Staff, RCAF: *After those hangars had been up for a while the wood would shrink a bit, and we used to spend days with a big wrench with a piece of rope tied to it so it couldn't fall on someone; we had to climb all through those rafters and tighten the nuts.*

Large numbers of construction workers were essential if schools were to be operational with a minimum of delay: engineers, carpenters, plumbers, and qualified supervisors and administrators with industrial backgrounds. Preparations for the arrival of the first aeroplanes continued through the winter of 1939-40, so that training operations could start on schedule in the spring.

Civilian Construction Helper: I worked on these different training schools when I was about fifteen and sixteen during the summer holidays—that would be about 1940. At one place I was assisting the engineer on surveying; he was an older guy and he couldn't see too well, so I used to take readings for him on the transit. The runways were in first; they'd just trim down the earth and then pack it and roll it—the buildings came later.

We had to get a converted steam roller to Estevan. The boiler had been cut down and we had a Mercury engine in it and we had to get it from Winnipeg to Estevan. You couldn't get any flat cars at that time, so the only thing to do was to drive it—and I drove it. It was at about four miles an hour, and I was over a week getting to Estevan.

Building Contractor: I was a young engineer at the time and we were the suppliers of the basic steel for reinforced concrete and anything to do with foundations. Buildings were going up all over the place; we were putting huge hangars and training facilities at Portage la Prairie and throughout Manitoba. We hardly knew what they were for and we didn't ask any questions. The government said, "This is where you're going to put the steam heating," and you could say, "What do you want the steam heating for?" And the answer would be, "Never mind; just get started."

Engineering Officer, RCAF: A great number of these stations could be made operative within six months. There were two aspects: the preparation of the actual airfield and the runways, and the construction of the buildings. So we had two groups of contractors and things proceeded as individual operations. Then there was the business of adequate oil and gasoline and the provision of storage tanks.

There was a constant pressure on all aspects of this thing. We would live in fear that because of circumstances in England—that they'd been bombed or something like that—that a group of 400 people would be arriving in Halifax five weeks early. Well, where in Hell were you going to put them? You could put up a building, but when it comes to the fall and winter that building had to be heated.

foundry that made the first threshing machines in Canada.

Bombing and Gunnery School Buildings of Standard Type

(July 20, 1940)

Fingal Air Training Project Will Correspond With Nine Others in Canada; Personnel of 70 Planes Expected.

Buildings under construction and projected for the Fingal Bombing and Gunnery School are of standard design and materials to lower costs and speed completion, according to information received from the Department of National Defence for Air. The six hangars are each 224 feet by 160 feet. The building area is 45 acres and the aerodrome under construction must be capable of 3,000 foot landing strips in four directions. Each runway must be 2,500 feet long and 160 feet wide. Bombing and gunnery schools are the largest type of school under the British Commonwealth Air Training Plan. Such schools have more planes, more buildings and more personnel than any other schools. Normal complement of aircraft at Fingal school will be 70, mostly the British single engine bomber-reconnaissance Fairey Battle with a crew of two, or the two-place North American Harvard, a low wing monoplane intermediate trainer.

Required at the bombing range,

Engineering Officer, RCAF: The whole building industry chipped in. Mr. Howe was the Minister of Supply and he was extraordinarily helpful—and although he insisted that things be done in an orderly way, there were all kinds of corners cut.

Building Maintenance, RCAF: Works and Buildings had a complete set of servicing personnel, electricians, carpenters, and the whole works. They had their own people on strength, but they would have a civilian establishment as well, regular plumbers and electricians and so on—just ordinary people that would be employed by the section, and they would carry out routine maintenance and minor repairs.

RCAF organizers foresaw the need for some 120 airports in a variety of sizes, locations, and configurations. Some would be satellite fields, having only a minimum of essential facilities, and the Department of Transport considered that 24 airports already in use could be adjusted to the requirements of the Plan with minor additions. This still meant that about eighty new airports would have to be carved from scratch.

Engineering Officer, RCAF: When we built this place at Clinton, the radio chap in the Royal Air Force—a Wing Commander—came over. He was an expert and he stayed with us at HQ for a while with the plans and saw that what was being done was in line. Then he was named the Commanding Officer of the station.

There were about 300 RAF people went over to start and to help operate this and to get our people trained. Of course, the place had been very hurriedly constructed, and there were things that weren't there that should have been there, and on the last afternoon I said to the Wing Commander, "If after you're there a while you find that you should have something that you haven't got, just give me a call and we'll see what we can do." About two weeks after that I had a call, and he said, "You people have done a jolly good job at this thing. It's absolutely astounding what's here in such a short time. But you know, honestly, there's not a bloody bath in the place—lots of damned little showers, but not a bloody bath!" He was a chap with a browser mustache—completely and utterly RAF. Well then, of course, we were able to fix him up in his quarters with a bath.

When I went there on an inspection trip about three weeks after that he said, "Oh, thanks for the bath—that made all the difference—but come and look at the NCOs' quarters." I went in, and some of the NCOs had gone downtown and bought those flat rubber stopper things that you can use over a sink drain. The showers had an edge around them about four or five inches high. They'd put these stoppers over the drains, and these were the baths for the NCOs—they'd just fill up the showers!

The Air Council

To be successful, an industry must first have a marketable product—the greater the demand for that product, the better the chances for the growth of that particular industry. At the head of every successful industry lies a management organization, and above management, a board of directors representing the shareholders.

In 1940 the product with the ready market was aircrew, and the British had placed an order for an almost unlimited quantity. The Canadian people were the shareholders who had put their trust in the Mackenzie King government as the Board of Directors to whom fell the responsibility of selecting a suitable management to see that Britain's order was filled on time. The industry was the Royal Canadian Air Force, and the management group was the Air Council.

Official, Dept. of Finance: Air Council dealt with the whole Air Force, but they had a sort of management committee for the British Commonwealth Air Training Plan, on which the British, the Australians, and the New Zealanders all had representatives. Canada's Department of Finance was also represented.

Officer, RCAF: Air Council was the body that had to make the decisions. These were the senior people in charge of the various activities—the Air Member for Training, the Air Member for Technical Services—and each was affected by decisions of Council which had to be passed down through the system. Senior personnel of the Air Force assigned to these jobs, along with the Minister and his Deputy, formed Air Council.

Officer, RCAF: The Chairman of Air Council was the Minister of National Defence for Air, which at one time was Chubby Power, and the Chairman of Air Members was the Chief of the Air Staff. Anything in the line of operations came under Air Staff, and that included Supply, Engineering, and Training, and these sections were all headed by senior officers in the RCAF.

The Minister had quite a group directly under him, and they were generally high-priced civilians—dollar-a-year men—and they worked with the appropriate Air Force groups. In a lot of cases they were more or less advisory, they were familiar with their own particular fields and they could get answers quickly, and the important thing was to get things done.

PR Officer, RCAF: A number of damn fine people sat on Air Council. The Air Force people were the heart of the whole matter, but they didn't know anything about how to *build* aeroplanes or parts and they didn't understand higher economics and how much money was going to be needed for a particular program. They hadn't been trained in industrial forecasting as had the head of the Ford Company or James Duncan of Massey-Harris who came in as a Deputy Minister for a while.

Almost all of these men were brought in on a dollar-a-year basis. One of them was a first-class lawyer, who went on to be the head of National Trust, and Conrad Black, a master accountant, came in from Winnipeg. And there was Macdonald of Macdonald-Currie came—there were some of the biggest names in the country in there.

Not all of them needed to stay; once the thing got going they were allowed to go back to their own businesses, but you had a collection there of some of the finest minds in Canada.

Assistant, Dept. of Munitions and Supply: The Department of M & S staff was recruited almost entirely from the business community. All the heads of the various production units and the control units were people who were brought in from outside, and they were called dollar-a-year men. Ralph Bell was the Director General of Aircraft Production, and Harry Carmichael, who had been president of General Motors, was the Production Chief.

PR Officer, RCAF: There were special meetings of one kind or another, and every once in a while, when public relations was discussed, we were asked to attend. You saw then ten or so senior Air Force officers and maybe eight very able, shrewd general industrialists or businessmen, or lawyers or politicians, and you saw the way these people could argue, sometimes fiercely, and no grudges were held. They'd finish a discussion and "Alright, then, this is the way we'll do it," and they'd either reach a compromise, or *not* reach a compromise, but whichever side carried the most authority would make the decision. There would be quite a normal dissimilarity between the way James Duncan would think about a subject and what an Air Commodore or an Air Marshal would think about the same subject.

It was done in a very impressive way. Decisions *had* to be made; the goal was so immediate and the need was so great for this Training Plan that all these men were big enough to have arguments—big, tough arguments—but to know "By the time we've finished this meeting we will have had to have reached a decision, and we've got to live together immediately afterwards."

Grant Dexter was with the *Winnipeg Free Press*. I called him at the Press Gallery one day and said, "Grant, I think it's time you came over and had a bath in this place." He said, "I'd love to."

The head of recruiting was there, the head of training, and the head of equipment, and they had their charts and they were telling him, "there's this and there's that"—all the needs, the things you couldn't get, but it was all going to come together at the right time and increase its production as it got larger. They talked very frankly and finally they stopped and they looked up and said, "Well, Mr. Dexter, what do you think?" Grant had a very serious mind, but a great sense of quick humour. He said, "I think it's great. Moreover, I think if all these things fall into place at the right time, it's not only going to be a great plan, but it's going to be the greatest coincidence in the history of man!"

Flying

Within a very few days of Britain's declaration of war on Germany, September 3, 1939, all 22 clubs of the Canadian Flying Clubs Association were involved in training new student pilots for the RCAF. Fortunately, in the preceding years the RCAF had encouraged the clubs to adopt RCAF training systems and standards. In November 1939, during the discussions with Britain, N.A. Seymour, President of the Association, had assured apprehensive negotiators that the clubs were capable of handling the expected flow of trainees.

In order to operate Elementary Flying Training Schools for the RCAF, each of the clubs was required to reorganize into a commercial company, and each of these companies was then required to post a bond to indicate good faith and stability. Also, they were required to satisfy the Canadian government of their ability to supply competent administrative staff along with sufficient qualified

NUMBER ONE

As a Canadian member of the RAF, Group Captain Robert Leckie had played a major role in the organization of the RCAF. In 1935 he was responsible for RAF Reserves and suggested then that Canada would be a most suitable location for training an expanding RAF force. Egypt and Cyprus had also been considered, but Canada was relatively close to the United Kingdom and its proximity to the United States would simplify the acquisition of aircraft, parts, and equipment. In 1940 Robert Leckie became a key man in the eventual success of the BCATP.

Officer, RCAF: Leckie was a Canadian in the Royal Flying Corps in the First War; afterwards he was with the Canadian Air Force. Later he went back to Britain and stayed there until coming once again to Canada in the spring of 1940 as Officer in Charge of the BCATP. He was purposely selected for the number one job as the Air Member for Training. I was one of his number twos. He was a very fine man to work for, just a top executive. He had all the patience in the world and the ability to draw people out and get the best from them. He spoke clearly and made himself understood.

Officer, RCAF: He was a small, alert, very keen individual; on the go all the time. An extraordinarily able man.

Officer, HQ Staff: Leckie was a little martinet, but he knew all about flying and all about the Air Force. He was a spark plug, a real ball of fire. When he got on a station he'd go through the place like shit through a duck, right or wrong!

flight instructors and ground personnel. Although early Elementary Flying Training Schools were operated by the clubs, commercial air operators were later called upon as airfields in their part of the country approached completion. In the very early stages of the Plan, the RCAF stations at Borden and Trenton were reorganized; Borden became No. 1 Service Flying Training School and Trenton took on the task of establishing a school for flying instructors, Central Flying School, Trenton.

Officer, RCAF: People who had been flying went to Trenton and were given Air Force ranks, and then used as instructors. A lot of bush pilots came in, and other fellows who had been flying as civilians with the clubs and so on. They then went to the staffs of the new Service Schools that were opening. From one station that was operating we would bring more people into the instructor ranks, then split the staff and start another. In this way we opened new schools which were Service oriented—SFTS.

Officer, RCAF: The schools fell into two categories in the way of management. The Flying Clubs Association took on the responsibility of running the Elementary Schools with Moths and that sort of thing. The instructors for these operations were essentially drawn from lads who had learned to fly in civil life. The Service Flying Training Schools and the Air Gunnery Schools were managed by the Air Force, primarily with RCAF officers but assisted by officers from the Royal Air Force after the Plan got under way. The Air Observer Schools for training navigators and bombaimers were organized and operated by already established civil air operators using Avro Anson aircraft. Initially these were flown by civilian pilots, but as time went by RCAF staff pilots were fed into the system.

Junior Officer, RCAF: Flying instructors working at clubs generally had another job, but they were instructing students on weekends. When the war came along and we started looking to the flying clubs as Elementary Schools that guy was suddenly working full time. So we had a nucleus there actually doing a job, to lead to a terrifically expanded job.

Officer, RCAF: The Elementary people had to have certain qualifications as instructors, and if the company required some to get special training they would be sent to Trenton, and then they'd go back to the company. Anybody that needed refresher flying or that sort of thing would spend some time at Trenton.

Officer, RCAF: In the very early years education was the big thing and the first students that the Air Force accepted were university graduates, but they certainly ran out of *them* in a hurry, and as the years went by the educational requirements were lowered.

Everything was put together in a hurry. There were not nearly enough instructors in Canada at that time to fully man any of the schools, so the initial graduates of the Service [advanced training] Schools became instructors at other schools. Early on almost the entire compliment of students became instructors, then later from every graduating class only so many went out as instructors.

Bush Pilot, Civilian: I had put in an application, and while working for Yukon Southern I got this letter from the RCAF to report to the Recruiting Office in Edmonton. I came in from Fort St. John and went down in the late morning and joined and had my interviews all on the one day. That same evening I was off to Trenton to Central Flying School.

There were three circuits at Trenton of different sizes, with faster aeroplanes on the outside. There were no runways, it was all grass and at times you couldn't find a clear place to line up on—you'd just figure the guy ahead of you would finish his roll and be out of your way in time. The traffic was very heavy and congested, and they weren't all pros either!

They had three special Advanced Flying courses in 1940 and there were about twenty men in each. On our course there were only about four Canadians—all the rest were Americans. We flew Yales that were meant for the French government, but France had already fallen, so the RCAF bought them—the instruments were all in metric.

We had a certain patter, an instructor's patter of different sequences that we had to learn, and it was all new to me. We'd put our helmets on—they had the old Gosport system where you'd speak through a tube to your student—and we'd sit on Trenton's stellar steps chanting this patter to get as proficient as we could before we had to go and perform before our instructor for our final check. It looked like kind of a nut house with all these guys sitting around talking to themselves, with a book in one hand, trying not to look at it, and speaking through these tubes with our flying helmets on.

Licenced Private Pilot: I had about 250 hours and my Private Licence. There was a real shortage of pilots, there were roughly 1200 licenced pilots in all of Canada in 1940, with any kind of licence, and there were very few guys in the RCAF with Wings then, so when I went into the Air Force as an Airman I was promoted to Sergeant the same day. I was over age to join—you were supposed to be under 25 and I was 28—but because I had built up my flying time the Air Force wanted me.

I went to Regina and joined the Elementary Flying School there as a student and went through in roughly two weeks. I flew Tiger Moths and Fleets, and the instructors were all civilians; actually they were Air Force pilots on leave without pay and weren't wearing any uniforms. From there I went to Camp Borden to take Advanced Flying. They rushed us through so fast there wasn't time for a Wings presentation. Then we went on to Trenton to get our Central Flying School Instructor's category.

When I left Trenton I got leave without pay from the Air Force, and then went to No. 5 EFTS in Lethbridge as a civilian employee. I was checked out in a Tiger Moth in about twenty minutes and told, "That's it; go to work." There were only five other instructors on the station, and we really flew like mad; we were going over 125 hours a month.

Officer Trainee, RAF: I was a Pilot Officer Provisional in the Royal Air Force at Calgary. I had just finished and passed all my tests and some Flight Lieutenant said that I could go on for my Service Training at Camp Borden—and then the war broke out on September 3rd. I was on the train on my way to Borden when Canada declared war on the tenth and I had been transferred to the Royal Canadian Air Force before I ever arrived there. They wouldn't send *any* pilots out of Canada then, so my RAF thing was finished, but I went on for my advanced training at No. 1 SFTS Borden, and that's where I got my Wings, just before Christmas in '39.

WE HAD TO SAVE OUR UNIFORMS FOR TAKING GIRLS OUT

Student Pilot, RCAF: We had four groups of five students in our class, and our group went to the last of the Flying Clubs. They hadn't all been converted to Elementary Schools, and the Edmonton and Northern Alberta Aero Club was still that. It was training RCAF pilots, but was still a club. They would send the odd instructor out to be trained in Air Force ways, but that still left others on the staff.

We had two Fleets and two Tigers and they belonged to the club. One Fleet had no coupe top, no tail wheel, and no brakes. It was as close as you could get to the World War I types; all I needed was a white silk scarf flying out behind. All these aircraft became RCAF when the EFTS was formed, but at this time they still had their civilian lettering.

We were *the* Air Force in Edmonton then—the five of us. You never walked anywhere, you always got picked up—a wonderful way of being entertained. We were issued with only two uniforms because of a short supply. We had one blue and one brown, so we never wore a uniform—we had to save them for taking girls out. We just went in our flying gear all day with old fatigues on.

We didn't have any Service aeroplanes come in; the only one would be the Examining Officer and he was an Air Force type. He came in and did our 25-hour check and our final 50-hour check. The only other aircraft that were using the field were bush planes coming in for their fall overhauls.

Our Fleet had a tendency for the engine to stop; the float would jam in the carburetor, and we'd land in a field and lift the cowling and tap it, and then try it. If it caught, you'd run like hell and grab the cockpit before the thing got going. Once we got in with horses and they wanted to come over and examine the aircraft, so I had to keep chasing them away and running back until finally the instructor said, "I'm gonna take off. If you can grab hold as I go by, OK—otherwise head for the road and I'll send a car out."

It was very frantic because they were going like stink to get us through and get us down to Trenton so that they could then get us overseas. But in the end they kept the majority of us back as instructors to start to fill up the system.

Officer Trainee, RCAF: After I left Trenton I went to St. Catharines as a Testing Officer. It was an Elementary Flying Training School that was just starting up and they were getting their nucleus of instructors. I was an official RCAF instructor at that time, and these boys came out of somewhere as civilian pilots and were passed out by myself and my supervisory staff as being qualified as EFTS instructors. There were a few of these fellows from the bush, but most of them had gone to flying clubs and put in 100 or 150 hours or so.

Student Pilot, RCAF: They had brought in all these bush pilots and had made them nominal Sergeant Pilots. But they never had a uniform and they never had a set of Wings till much later in the war.

Officer, RCAF: They went out and passed a flying test and if they were good enough the Chief Flying Instructor hired them. He was a civilian and he wore a Prairie Admiral type of thing—a uniform of sorts—but all the flying instructors were civilians. They had their Private Pilot Licence and quite a few of them had a Commerical Licence, but I gave them a patter test—it was pretty simple.

At those Elementary Schools they had no people with proficiency enough in Air Force flying to be the flying administrative people. But the civilians provided the staff for Air Force students

37

to be trained by civilian instructors—and then to be checked out by Air Force personnel—so that they could go on for their Service Training or be scrubbed from pilot training completely.

I not only had to test the proficiency of the civilian instructors, but I had to check their students when they were supposed to be ready for solo. It was my job to say they either fly or they don't fly.

New stations were opened at a rapid rate, but before any flying could be carried out the facilities had to be prepared by RCAF ground personnel, and generally these were recent graduates from the Technical Training School at St. Thomas.

Ground Staff, RCAF: From St. Thomas we opened up Aylmer, which was a Service Flying Training School. I was sent down as one of the advance party, and it was pretty raw. They had just finished it and all the dormitories were there, all the hangars were up, the runways were in, and they were just getting ready to bring in the first draft of pilot trainees. The station was equipped with Harvards and Yales and they were all in the hangars. Then they brought in their first people, who had just graduated from Elementary Schools at different places. Early on in the game we had a lot of Australians at Aylmer.

Not all the stations were as well organized. During the start-up phase of the Plan, there was so much work going on at such a large number of stations across the country, that the first Air Force crews would arrive before the contractors had completed their work. It was often necessary to appeal to the trainees' pioneering instincts.

Warrant Officer, RCAF: The hangars were not completed and the quarters were not completed—and there was no messing. There were a few supply people and two or three men, but no officers. The next night there were some eighty airmen came in from St. Thomas, so I had to put them up at hotels and the YMCA, and I had them eating in restaurants. This went on for some ten days before we had an officer on the station, so as a Warrant Officer I was the CO until the officers arrived.

Recruit, Women's Division, RCAF: At Dunnville, the station had just been opened and they had a lot of work to do. We'd go into the barracks wearing rubbers over our shoes, but the rubbers got stuck in the mud—you were just covered with it. Then the water in the taps was black, so every night an airman would come with a huge pan of milk, and we used this to clean our teeth and try to wash up a bit.

Officer, RCAF: They got an old cook stove somewhere, hired a cook, brought the airmen onto the station, and started to feed them. When they turned the water on the lines were full of white lead and all the other things, and the crews were quite ill with dysentery for three or four days. It was a bit rough. There were no washroom facilities; you just had a little shack in the back with a two-by-four and everybody sat over the two-by-four—if they could make it—but a lot of the fellows never made the little shack and they hung out the barrack's windows. It was a bit messy.

Ground School Instructor, RCAF: They scrambled for personnel at the start. We had a nucleus of staff, but they hired people from all the surrounding towns and used them in the kitchen and hangars and so on, general duties. But it gave them quite a break to have some work in poor Saskatchewan, which hadn't seen rain for ten years until we got there—and then it rained like hell.

Joining

As BCATP schools opened all across the country and the RCAF set up a system of recruiting stations, it became obvious that Prime Minister King was correct in his belief that "the Air Force would never be in need of conscription." The youth of Canada had been interested in aviation for many years.

I was eighteen, and everybody wanted to join up; you didn't want to be hanging around, especially if you came from a small town and all the other guys had gone. They had a big thing about attitude. Everybody wanted to be a fighter pilot, and so you were terribly conscious that you had to make the right impression and that you had to be "keen". All the guys would get together and rehearse what the right answers were to all these stock questions like, "Why do you want to join the Air Force?" Well, the consensus was that the right answer was that you wanted to join "because *somebody* had to do *something* to stop the scourge of Nazism," or something. And this sort of idea was, of course, complete bullshit.

I joined in Ottawa when I was seventeen and a half. My father was overseas at the time and my mother said if I got my Grade 12 that she would let me join the Air Force. All my friends were going at that time and I had been bugging her. So in June of '43 she let me go.

I didn't enlist right away because it didn't seem like it was going to be a very long war. Everybody figured, "Oh, this thing's gonna be over in a few weeks." So I sort of held back because I wasn't interested in joining the Army, and when I did come around to enlisting it was very difficult to get into the Air Force. There were so many volunteers for aircrew it was standing room only, and there was a long waiting list.

I graduated in 1932, and when war broke I was manager of a drugstore in Saskatoon. I was too old for aircrew then and I had to wait for over a year before they accepted me, and when I finally got in most of my buddies had been knocked off overseas. By that time I was thirty and most of the kids were under twenty. I was like an old man, and they all called me Pop.

One time I went out around Cartierville and I saw these light aircraft and a big sign: "Fly five minutes for two dollars". So a friend and I got up our courage and had a five-minute tour around the airport, and thought it was just great. The thing that amazed me was how young the pilot was. I expected some older man—a more mature person—but this fellow wasn't more than about his mid-twenties, and we were in our late teens. I sort of thought, "Well, goddammit, if he can do it, why can't I?"

When I was wiring Handley Page bombers, putting in switches for the bombsight and things, I sometimes used to sit there at that little table and think, "I wonder what it's like to look out that window and press the button?" Technically, I was an electrical apprentice, but in those days you worked anywhere the company sent you for the fantastic amount of twenty cents an hour, sixty hours a week, and no overtime. I hand-riveted ribs for the Fleet Finches, I helped assemble Fairey Battles, and I worked on the fuselages of Handley Page Hampdens that were built at Fleet at Fort Erie and sent to Malton for final assembly.

They weren't taking any boy soldiers or buglers or anything like that into the Air Force, and I was more or less resigned that I would have to grow up and get to be eighteen or nineteen to go. I was wiring the front panel of a Handley Page and the noon hooter went, and I simply dropped the navigator's table which was folded up into the side and put my lunch bucket on it. A friend of mine came along and said, "Hey, there's a mobile recruiting unit down at the Armouries for the RCAF. We're all goin' down." I said, "What am I going to go down for; I've got Grade 9 education and I've got two years in trade school, which is really no academics." And I was only about five feet tall and I think I hit 98 pounds. "Well," he said, "get the foreman to give you a slip. If the recruiting thing stamps it you've got the afternoon off—with pay." So what the Hell did I have to lose?

I ran down to the Armouries, and this great mob of guys was there. I tried to look older—you know, I think I even tried to smoke. They put us in blobs and they sat us down and gave us an IQ test, which I evidently passed, because everybody that didn't was put on one side and they talked to them about something else. Finally there was about forty of us left, and then a fellow came in—I guess he was the Sergeant. He said we'd all made P or O. I wondered what P or O was—it meant pilot or observer. Well, I didn't want to be any pilot, but he said, "You passed all this, and if you pass the medical you could go into training as P or O, but you've only got Grade 9 so you'll have to go to the War Emergency Training Plan." Which turned out to be ten bucks a week and pay your own room and board for three months in Hamilton. I still wasn't eighteen.

I lived at Mother McCurley's. She had ten or fifteen of us there—two to a bed. She gave us breakfast, she gave us supper, and she gave us peanut butter and jelly sandwiches for lunch, and we went and learned mathematics and all these things. It seems to me that there was an exam every Friday and a big exam on the fourth Friday. You could fail the in-between Fridays and catch up on the others, but if you failed the fourth week or the eighth week they came and talked you into being an air gunner or a wireless operator or something.

Mother McCurley was a tough old gal. I remember once asking her for more food at suppertime. She packed my bag and stuck it out on the front veranda. She said if I didn't like what she was doing, I could leave. But Hell, you couldn't leave her. I only had a buck and a half a week to spend on laundry and you name it.

At the end of three months they marched us down to a train, and the train came in to the back end of Manning Depot in Toronto, and there we were.

THE GROUND STAFF

Junior Officer, RAF: *We had to keep those aeroplanes flying and we had a stiff schedule of inspections to meet. So we pressed the guys harder and harder, till one day . . . down tools. They said, "We're quitting. That's it!" So I was faced with a mutiny. They were having to cut corners and they weren't doing as good a job as they would have liked to have done. These guys were trained there's no substitute for quality. They had to do the thing right, which of course was the proper attitude. But we got it sorted out and they settled back.*

HOLY HECK... I COULD FLY AS GOOD AS THEM!

Ground Staff, RCAF: I was transferred to No. 8 Repair Depot in 1941. It was a strip of concrete about half a mile long which ran down the centre of hangars facing onto it, like rooms on a corridor in a hotel, and the doors opened onto the concrete.

The hangars were designed for certain jobs. One was engine repair, another was airframe repair, and metal working shops and various maintenance areas. Any of the aircraft in the Commonwealth Plan that were damaged beyond base capability would be sent to the repair depot, flown in if possible or brought in by truck. And then they would do a major repair or an engine overhaul or whatever.

There were Harvards and Ansons and Yales and the old Fairey Battle with the big tapered wing—there were all sorts of aircraft coming in all the time. And then after the engines had been overhauled they'd mount them on test beds and run them up.

Station CO, RCAF: On the Elementary stations virtually all the ground personnel were civilians, and without them we wouldn't have been in the air. Our flight line foreman was an absolute wizard; he kept an eye on all the maintenance personally, and you were always sure that all the things were done that were supposed to be done.

Ground Staff, RCAF: I wanted to get into aircrew right away, but they had these quotas, and the aircrew quota was full. So I thought, "Well, I'll take a ground job," just to get in, thinking once you're in it would be easy to get into aircrew. What a dreamer I was—it took me a year and a half.

Ground Staff, RCAF: I watched all these take-offs and landings, and I was a little bit on the cocky side, and I thought, "Holy heck, all these kids and some of them not as old as I am. I could fly as good as them if I just had a chance. The Flight Commander heard about me and said, "OK, get into the front of that Harvard with a chute." So I did, and he said, "All right Corporal, I'll taxi it over to the runway for you." He lined it up and said, "You have control!" Well, about the last coherent thought I had was when I opened the throttle.

In the spring when the snow was melting, we still had to get those aeroplanes with skis into the hangar. They came up with a wheel with an arm on the axle that fit into a hole on the pedestal. It took a little more muscle to push, and there were times when we'd start them up to get them across a piece of tarmac.

They were short of mechanics during the early days. There was a little place outside Borden where they had a field, and I would go out there with seven students, seven aircraft, and an instructor, and we'd go back and forth by truck. There'd be mornings when I'd have all seven aircraft running up on the line with no one in them; there just weren't enough mechanics to go around.

Before the war started Norway had ordered Mohawk P-36 fighter aircraft from the United States, and they were on their way to Norway at the beginning of the war, but they managed to divert them. I was sent to Curtiss Wright in Buffalo to learn all about the Mohawk. We started getting those in to Little Norway just before Christmas of 1940.

I was interested in aeronautics when I started high school. I was always interested in a trade, because my father was a blacksmith and a carriage maker, and I always worked in the shop with him from the time I was a kid. In high school I wasn't interested in the academic subjects, but was at home in the trades. At Ottawa Tech I specialized in aeronautics, and then when the war broke out I thought, "I'll get into the Air Force as a mechanic."

Student Navigator: *For bombing they used eleven-pound practice bombs that were hung on a rack, and they'd just pull a toggle to let them go.*

Ground Staff, industry: *During the first part of the war the AOS were using the old Ansons I through IV. They were all tubular steel and fabric-covered structure in the fuselage, and the wings and the tailplanes were all wooden. They all had bomb doors, but there was a variety of different equipment in them for different roles, and most had the Cheetah IX engine. You started those engines right from the nacelle, standing on the ground. You'd turn the inertia starter mechanism, which you wound up until it got to quite a high speed, and then you engaged a geared starter into the engine, and that would get her going. There was no electric starter on those early Ansons.*

Engineering Officer, RAF: *Those darned Ansons weren't really made for maintenance; they weren't made to come apart. If you wanted to change a wing you had to take the engines off and the undercarriage off. You had to take the whole thing right to pieces to put a new wing in.*

Ground Staff, civilian: *At the Elementary Schools we used to have to swing the Moths and the Fleets by hand. Sometimes when they were hot they'd kick back, and you'd have to watch them or they'd take your hand right off.*

Ground Staff, RCAF: *The old Fairey Battles were an all-metal aeroplane, and I guess pretty modern for the time; they were built like a brick shithouse. The only real problem that we had with them was that the undercarriage would sort of hang up, and the guys would have to dive them and get up some speed and try to shake them down. But even if they came in on the belly, it wouldn't hurt them too much.*

Ground Staff, RAF: *We had sixty Harvards at Medicine Hat. We had four hangars, and one hangar would take six Ansons and then you could put two or three Harvards in as well, but it was difficult to operate in the winter time. We had no way of using a tractor to pull the aircraft out—we didn't have any towing gear—so we used to get half a dozen guys on either side and push.*

Ground Staff, RCAF: *A staff pilot was going to take off in one of the old Lysanders, and I yelled at him that it didn't sound too good, but he went off anyway. He got to about 200 feet, and it just quit dead. He made it into a field OK, but it was in the spring and the ground was soft, and the old girl put her tail in the air and rolled onto her back just as nice and slow as you could ever want. The guy was kind of mad, because he had to climb out into a big puddle.*

Ground Staff, civilian: *At night after the flying was through for the day, the aircraft were all pushed into the hangar. We could get a couple of dozen Moths in one of the single ones. In the morning we pushed them out and lined them up and made sure they were all gassed and oiled. Then we'd run them up for a good fifteen minutes and check the mags, and by that time the students and instructors had arrived.*

2

The terrors of Manning Pool
Guarding everything
ITS and a whole new ball game
The thrills of Elementary
They had that big brother image
Three Hail Marys and the first solo
Getting on skis was different
Little flares all in a row
The complications of aerobatics
Those bothersome coupe tops
It was a matter of washing out

THE FLEDGLINGS

Manning Pools

It all started at the Manning Pools: that first long night away from home; the first taste of Air Force food; the first uniform; the first Drill Sergeant. For some it was just a new experience, for others it was a job, but for most of the recruits it was a shock.

None of us had any money in those days and I said, "What should I take when I go to Brandon?" They said, "Take a toothbrush but nothing else. You'll go down there, they'll pay you, they'll clothe you, they'll feed you. You don't need anything!" I was naive enough to believe them, so with three or four dollars in my pocket and a ticket to Brandon I got on the train. After a couple of days on the train and no change of clothes, I got to the Brandon Depot and found that they didn't have any clothes for us. I had to wash my socks out at night and put them on still wet in the mornings.

Finally I got outfitted. I wore a 7½ shoe and I got 9½ boots—the toes curled up just like the RAF caricatures. The pants were wrinkled and the brass buttons on the tunic were yellowed. I was with a fellow from Calgary and we were out walking. I really was very pleased to be out in my new blues, but a sergeant stopped us in the middle of the street and he told me that I was a real liability to the Air Force and I was to report back to barracks.

I arrived in Saskatoon on December 6th, '41—about thirty below zero and with no coat—freezing to death. When I did get one it was a new issue and it had five buttons down the front instead of three or whatever—a very smart-looking greatcoat. And I was fortunate, I was a size off the shelf and it fitted beautifully. The first morning on parade the CO came along and he stopped—and of course we're just Acting AC 2, the lowest of the low—and he said, "Where did you get that greatcoat?" I said, "I was issued that coat." He said, "Take it off; it's non-issue!" And I said, "But you just gave it to me!" He turned to the Warrant Officer and he said, "Where did this airman get this coat?" and he said, "That's what we're issuing now, Sir." And the CO said, "Far too good for them!"

Manning Depot in Toronto was the bull pen at the exhibition grounds. It was that nice barnyard smell down there in July. We put on those heavy blues, and those heavy shirts, and then the heavy blue tunic over the top, and, oh, boy, it was like a roaster in there. Instead of drilling us for fifty minutes and resting for ten, they used to drill us for ten minutes and rest us fifty.

We were thrown these rough uniforms that didn't fit, and we were walking along in Toronto and saw a Flying Officer with his "Officer's Air Force Blue". My friend went up to him and said, "Gees, where'd you get *that*? Look what they issued me!" —right up to him—and he reached out and *felt* it. The F/O practically blew his stack. He was gonna put us on charge. We were so dumb we saluted the doorman at the Royal York—we thought he was a Group Captain.

We had no equipment. There was nothing there but bunks and blankets. We got our shots and the guys fell like ten pins. I had an upper bunk assigned to me and I remember getting my arms on the bunk but not being able to press down hard enough to flip myself up—somebody had to lift me. Moanings and groanings at great length that night.

I'd never had a vaccination or inoculation and we were just lined up like a bunch of cattle and went through and got punctured. The next day we went through and got punctured some more.

Manning Depot was being inaugurated in Air Force ways—physical training, disciplinary drill, and that type of thing—to make a military person out of you. It was rough coming out of civil life; they had us up at the crack of dawn out running around the cinder track, and a lot of us were in white shoes. I guess the old drill sergeants really enjoyed running us into the ground, because they sure gave us a going over, but by the end of that training we were in great shape—I think the best condition I've ever been in in my life.

By the time we left we were pretty good soldiers. We could march with the best of them. We could take a Flight ourselves, say forty or fifty men, and give orders to have them form left on the right and that sort of thing. We had learned really in that short time how to handle a parade square pretty well. You could march them around and have them stop where they were supposed to stop, and start when they were supposed to start.

We had a bunch of corporals that terrorized us. One chap in particular—his nickname was Cowboy—revelled in just scaring the Hell out of us young punks. With the old uniform you had to take your boots off to take your pants off, so all day long it was taking your boots off and getting rid of your fatigues, getting on your blues, and putting on your boots. We seemed to be taking off and putting on boots all the time.

They had us doing everything: working in the canteen; peeling potatoes; cleaning latrines; serving in the Officers' Mess; and all the other joe jobs. Cleaning barracks; painting white rocks out on the walkways; just about anything you can imagine.

The sergeant came along one day and said, "Who can type?" It was forty below—God it was cold—and the fellow standing beside me said, "Come on, we can type. Come on." And the next fellow said, "Yeah, you'll be unloading a whole carload of typewriters." Anyway, we volunteered to type and we wound up writing wills for all the fellows coming in on the drafts. That's the last time we marched, while everyone else was slugging it out on the parade square. That was a cushy deal.

Two of the senior NCOs were ex-Winnipeg policemen who used to frequent the drugstore that I jerked sodas in through '38 and '39. We fed them doughnuts and coffee for years, so when they spotted me on the parade square, I guess they thought, "There's a guy that knows how to work behind a bar." I got duty in the Sergeants' Mess, which got me away from a lot of parade square drill and a lot of other tedious tasks . . . and I had a rather nice few weeks.

We were late reporting. The Air Force frowns on such things and each of us was put on charge. We lost pay and we were detailed to various jobs. We were CB—confined to barracks—and we couldn't go anywhere in Toronto for the first week.

We got our uniforms and we thought, "They're not gonna give us a *rough* job if we report dressed up." So we went down in our best uniforms . . . and I got detailed to where they wash up the pans for the kitchen. I forget how many guys were at No. 1 Manning Depot in Toronto, but there were pans from the breakfast bacon that just about reached the ceiling, and me and another guy spent the evening there washing out bacon pans in our best blues. But it was sort of the luck of the draw, because another guy got detailed to work in the canteen with all the YMCA girls and he was really sitting on top of the world.

I broke more eggs there in one day than I ever saw in all my life. I was assigned to kitchen duty and they were making cake and they were breaking eggs into huge metal mixing bowls—a hundred eggs to a bowl. That was the size of the recipe for making cake for 7,000 people. The things were immense; they were like small cement mixers.

In Edmonton we slept in the old horse barns, but they were putting a new heating system into the Arena. Those were the days when a good salary was a hundred dollars a month, and here was this new heating system going in. I asked the contractor, "How much are you spending on this?" And I can remember him saying, "$28,000." That sounded like a lot of money for a heating system.

At one stage I had about six days until pay day, and I had sixty cents. So I allowed myself ten cents a day for an ice-cream cone. That was my big treat for the day.

We went and sat out in the horse ring under our initials. They called all the As and the Bs, and I finally got my first pay. Some paymaster peeled out about $18 in cash, and it was all mine. I took it and of course gave him the big five and marched away. I never had so much money in my life and it was all mine. I didn't owe anybody a cent.

The streetcars used to come into the exhibition grounds in Toronto, and in those days you could buy a streetcar ticket for the whole day. The airmen used to wear it in their hat—very flashy, macho guys. So the first night I was allowed out I got the streetcar and somehow on a Saturday night I got to the corner of King and Yonge with eighteen bucks in my pocket. I walked up Yonge Street to Bloor, crossed over, and walked back again. Small town boy in the city. Yonge Street was lit up and I thought I was about four feet off the cement with those eighteen bucks.

I got back on the streetcar and went back to the Depot, and the other fellows hadn't left yet. "Hey, how's it out in the town?" "Oh," I said, "Noisy, lots of lights!" But I didn't drink, and I didn't smoke, and I didn't spend my money. Boy, talk about being a rube!

A delightful chap that I joined up with had led a very sheltered life and quite openly said, "The only reason that I got into this was that everybody at the 'Winter Club' was wearing a uniform, and I thought I should get into a uniform." He was naive to the extent that when he got off the train in Toronto he asked the corporal to carry his bags, because that was the life he had been used to.

He thought we'd be two to a room when we got to Manning Pool, and when his dad was coming down to visit him he was very concerned. He said, "I like the gang, but if my dad finds out I'm living with 4,000 people in double bunks, he'll get me out. My letters are leading him to believe that I've got a billet with one other chap." So he managed

I remember most the pay days. You didn't have a dime in your pocket when you joined up, and then suddenly you got your pay, which was not very much. But everything was provided: your food, your lodging, and your clothing. Then suddenly all this money came your way on pay day, and it was quite a strange feeling to a lot of us who had gone through the depression years.

to get a 48-hour pass and he and his dad stayed at the Royal York, and his dad never *was* exposed to the terrors of Manning Pool.

It was the first time in my life I ever got drunk. We were sitting in the cafe on the corner of Rosser and 10th, and we were drinking rye out of a brown paper parcel, and I'd never gotten into the juice in my life before. I can remember somebody saying, "Gees, he's a nice guy, but why does he drink so much?" I suffered for it the next morning, because I had to go on Church parade and they just held me up, but I sure as Hell was on parade because I was scared of the sergeant.

Guard Duty

Manning Depot may have terrorized some of the young recruits, and to others it may have appeared to be a useless exercise, but for many the next step toward becoming an accomplished airman was even worse. They were required to guard Air Force installations all across the country, and few had expected this phase of their training.

The thing was goddam stupid! They had this equipment depot and they had us marching up and down *inside*—guarding a fence. They had the open side unguarded, wide open to the prairie—just goddam stupid!

We did our guard duty at Rockcliffe, and this was my introduction to Eastern Canada. I found it very cold and uncomfortable. There was a station on top of one of the hangars and there was a station on top of something else. There were five or six different towers that we manned and a telephone system between them and back to the main guard house. We could always tell when the guard house was listening in because they had a big clock there and you could hear it on the phone: tic, toc, tic, toc. We were told, "Don't go doing a lot of talking. You should always be on the lookout!" But on the lookout for what? In the middle of the night you'd look out across the Ottawa River and wonder who the Hell would ever be coming across there anyway.

There was a lot of talk about German saboteurs, and Rockcliffe was one of the old key airports in Canada. We were guarding all the hangars and the wireless station and all the main buildings. We would move along from one post to the next and we would be on shift ten to twelve hours at a time. It was very scarey. We were green and we had rifles and live ammunition. You'd be standing in the dark and you could hear every little sound, and you thought someone was sneaking up on you with a bomb.

We were down in Dartmouth over Christmas and I was close to tears. The fog horns were going, and at midnight for New Year's all the ships blew their whistles. And I thought, "What the Hell am I doing out here? Waiting for German troops to start landing?" It was a lonely life. You didn't see anybody. You worked four hours on and eight off, and if the other guys weren't with you they'd be sleeping. You were really isolated. It was a bad Christmas and New Year's. I was twenty, and very much homesick. That was my first time out of Winnipeg.

I was on guard duty at the entrance to the camp one day, when this car came up with a flag flying from the fender which I didn't pay much attention to. I just stood there, gave a butt salute, looked in, and this chap had all this gold braid on his hat. He was an older gentleman; he looked like a very kindly old man. He kinda looked out rather intrigued by this raw recruit that had given him a butt salute, and he leaned forward and looked at me. So I leaned forward and looked at him, and kinda gave him a little tip of my hat, like an informal conventional salute, and nodded and smiled. And the old gentleman just kind of smiled and nodded and went on. The sergeant in the gate house nearly died a thousand deaths.

I hung around the hangar area whenever I wasn't standing guard with the hope that I could talk one of the pilots into taking me up for a ride, and finally my wish came true. A pilot said if I wanted to go I would have to sit in the upper turret, and I wasn't to talk, and I wasn't to move. I was to sit there until we got back. So I jumped at the opportunity and went off on this long harbour entrance patrol which lasted close to three hours. It seemed long, and it kept getting longer because I wasn't properly dressed for the occasion and I was absolutely frozen. It came to me as quite a surprise just how cold an aeroplane could be.

We were instructed on how we should challenge somebody. They had very remote spots around the airport, some of them right off in the boondocks, and they would drive you out in a truck and drop you off at your little guard hut. You'd be there for a period of two or three hours before being relieved. We had been instructed that if you were ever in a position where you had to challenge somebody you would call out, "Halt! Who goes there?" And if you didn't get a reply, you would say, "Halt, or I fire!" And if the person didn't stop then you would have to challenge him by actually firing your gun.

Part of the post was guarding *inside* the hangars, and all the aircraft would be stored away in there. There was every type you could imagine and sometimes I'd lean my rifle up against a wall and get into one of them and "play aeroplane". It was great. There were Ansons and Lockheeds and DC-3s, right down to the little ones.

We were 35 Yanks and about 35 Canadians, and the Yanks were full of pranks. They were getting awful fed up with this guard duty, so one night they started letting off volleys of shots, all around from the pitch dark. Our Guard Officer came around and doubled the guard, then he tripled the guard, then he put on a roving patrol of about ten guys, and they had to march around the perimeter all night. I was afraid we were all going to be washed out.

At the Queen Mary Road Station in Montreal we used to have a little hut at the gate where one guy stood with a rifle. You were on four hours and off four hours, and you got so you didn't know whether you should be sleeping or walking around, and every now and again you'd hear a noise, especially at night, and you'd get the rifle up. Of course it didn't mean anything 'cause it didn't have any shells in it. I was standing out on duty this night. It was snowing, and about four o'clock in the morning, and these four guys came rolling along towards the gate. Before I could say, "Who goes there?" *they* said, "Who's there?" I wasn't really awake and they took the initiative, and I told *them* who *I* was! And they said, "Advance and be recognized!" And here I was with a rifle walking out to meet these guys and *I'm* supposed to be challenging *them*! The bunch of us just sat down right there in the snow, and laughed and laughed and laughed!

In December we were all posted to Debert, Nova Scotia, guarding the airport that was being built. That's the A-hole of the world at that time of the year. Oh, it's cold and miserable—and 52 of us livin' and eatin' and drinkin' and fartin', in one little barracks. All AC-2 punks. We called ourselves the Fuck Fifty Two.

We were two on and four off. You'd get so bored that you'd hack away with your bayonet on

your sentry post. You'd start with your initials, but everybody was whippin' around it—anything to decrease the boredom standing there for two hours in the snow—and pretty soon it was just a pile of kindling. Three years at university and sittin' there waitin' to fly, and we never saw an aeroplane. At Christmas I was on duty, and everybody got a bottle of beer with dinner, but by the time I got off duty somebody had taken mine.

The girls around there were just so-so. We were never invited into a home. My sister back home was in nursing and she was introducing all these nurses to the Australians and the English guys, and they were all enjoying it, and we were sittin' in Debert. There were so many of us that it was hard even to *meet* a girl. One fellow was shakin' up on all his days off and he had our tongues hangin' out —"Such a set-up!"—and we were starvin' for females. But when we were posted to ITS—Initial Training School—he came down with a dose and had to stay in the hospital. He had to wait until he was cleaned up so we all thought, "Gees, glad we remained pure!"

Initial Training School

Once the boredom and discomforts of guard duty had passed, the students faced another challenge on the road toward their goal of becoming airmen.

The RCAF had arranged to take over a number of large buildings in different parts of Canada that were suitable for handling large groups of students for instruction in ground school subjects. Essentially these courses would provide a good basic knowledge of airmanship, but the syllabus was demanding and often the subjects required an academic background beyond the limits of high school graduates. For many, the real work *that was required to become airmen only became evident upon their posting to Initial Training School.*

This corporal had been riding me unmercifully; oh, God, I never did anything right. We'd just got our uniforms, and they didn't fit very well. Billy Bishop, who'd been my idol from the First War, was coming in the front door, and this corporal was in the entourage, and I happened to be "on post" there. They'd just taught us to present arms, and the bolt came out of this old rifle and caught my greatcoat, and when I brought the rifle up the greatcoat was like a tent! I'm embarrassed as hell, and this corporal, he's fuming in the back—he's gonna kill me! But Billy Bishop walked right up to me and he said, "You're doing fine, son. Thank you very much." Billy Bishop, my boyhood idol. He knew I was embarrassed. I'll never forget that.

THE I.T.S. OF CANADA

NO. 1. INITIAL TRAINING SCHOOL

No 1 ITS at the Hunt Club was *the* ITS of Canada. It was quite a posh place, and Toronto rich people had horses there before the Air Force took it over; the barracks were in the stables. One of the caretakers used to say, "The recruits aren't like they used to be. When the rich Americans came up here in the early part of the war they'd park their Cadillacs across the road and leave them there when they went off with the Air Force."

I went to ITS in Victoriaville. It was a Catholic convent, and we sat in the church and the chaps lectured from the pulpit. We lived right in the convent building. A lot of the boys were kinda roughed up, especially if they were in ones or twos coming back late at night. The French Canadian people wanted our money, but you couldn't date any of their girls; they just didn't want them to associate with us. We didn't mind, because we were worked so hard and we wanted to pass all the tests so we could qualify for aircrew.

The thing I liked best about ITS was the Link trainer; that's the first time we got into anything that even approached an aircraft. The walls of the room were painted with a horizon, with the ground below and the sky. It was strictly visual, but they gave you headings to fly and rate one turns and rough air and so on.

It was like a concave mural on the wall, and you had the impression that you were flying. After a couple of hours they closed the hood and then you were supposed to try to fly by instruments. The cockpit was about the same as we had later in the Tiger Moths.

It was very unstable; you could not centre the controls perfectly—it would always drop a wing. They'd tell you to turn around to the water tower or whatever on the horizon and stop there. So you'd put on your rudder and go 'round, and you'd end up jockeying trying to get it stopped.

The Link could be a great source of airsickness, because that was your first real experience in flying. It was only a simulated situation, but when the instructor would start to throw that thing around it was like a bucking bronco. There were many airmen who got violently sick, and that usually spelled their doom for pilot training.

We left our guard duty and were posted to ITS in Toronto, which was the Toronto Hunt Club on Avenue Road. We had our quarters and our meals right there, and we had basic courses in theory of flight and some Link Trainer exercises. The idea seemed to be to start weeding people out who weren't going to measure up to aircrew standards.

We got out on the odd pass to see a movie or that sort of thing, and we had route marches down the street past one of the girls' boarding schools. That always gave us a bit of a lift, because we'd see these cute little girls standing out watching us and waving—and that was a thrill.

It was everything about theory of flight. We all wanted to be pilots, and they taught us all kinds of things to get us ready for elementary flying. We were really glad to be there, because you're talking about aeroplanes and looking at pictures, instead of standing out on guard duty and feeling sorry for yourself. Now you're all dreaming about flying and being aces, and stuff like that.

I was four years out of school and suddenly realized that studying was going to be a whole new ball game, and I worked harder at ITS than I ever worked at any subjects that I ever had to take. I was terrified that I would not pass the ground instruction. Most of us feared the algebra and other mathematics, because we got into logarithms and trigonometry, which was something that in Grade Eleven in Manitoba you didn't get into. And so I was paralyzed with fear that "Boy, I'm not gonna make this."

We had a chap on our course who was one of the very few mathematical geniuses I've ever met. He was pumping gas in some little jerkwater town in BC when he joined up, but in those days a job was a job. The instructors hated him because he was always so far ahead of them that he was an embarrassment, and they told him, "Just keep quiet." At night we went into the classroom and this guy would teach us. He would pound the blackboard just like a school teacher, only better. He'd pound it so the place would be full of white dust, and his language was terrible: "Don't you bastards see that . . . " There were half a dozen of us that had a tough time, but he hammered it into us. He was so damned smart that he couldn't understand how anybody could be so damn stupid, but he got us through. He was shattered at the end of ITS when they put him as a navigator.

I recall these long tables in the billets. There'd be fifteen or twenty of us around the tables and we'd be studying and doing our homework. There were two people who were mathematical whizzes and they just walked around those tables night after night and taught the rest of us the mathematics that we couldn't pick up during the day. More than anything else I credit these guys with getting fellows through the math part of their ground school training.

I got the highest mark in mathematics through the navigational subjects that I ever had in my life, and for a very short period I lived in fear: "My God, maybe I made a mistake. They might send me to navigation school." If, in your own mind, you were headed to be a pilot, being a navigator could be a problem, and they might decide that your ground school combined with your other programs showed that you should go to a gunnery school or something else. What we didn't realize was that the decisions that determined whether you became a pilot, a navigator or an air gunner, was in fact the need for these trades. Many young airmen at ITS went to gunnery school because that was what the system needed at that time. And six months later there might have been a need for navigators, and later still they'd have a tremendous need for pilots.

I kinda came in over the fence about six in the morning and just got into bed with my clothes on before they called everybody to get up, and that was the day we were going down for medicals. The first thing they did in those medicals was to take all your clothes away and you'd all sit around waiting for different tests. Nothing was sacred in the Air Force. I was told that my eyes were "rather lazy", which I kind of understood after the night before, but I couldn't admit to having been out all night. Then rumours started developing that some of us were going out as navigators. I was kind of heartbroken, because I really was set on becoming a pilot.

The split came when you were down to the end of Initial Training. It was then you were made into pilots or navigators or bombaimers, and some went as air gunners. Selection was based on how well you did during the training. If you were a whiz at navigation, then you destined yourself to be a navigator.

We wore a little "fore and aft" cap, and as aircrew we had a little piece of white flannel that we put at the front that indicated "training for aircrew". The ground staff didn't wear that marking. There were different stories that went around about the flash, and one was that it was to warn the girls that you were married.

Before you went to a flying station you were allowed to put up the white flash, and you were proud of it because of what it stood for. When you went out in the evenings you usually shot the girls a pretty good line. But the boys in the other services, the Army and the Navy, they used to tell the girls, "You have to be careful of them because that white flash means they have VD."

There was no selection in any sense that I was aware of; there were never any interviews. In academic standing the top third became observers, the next third became pilots, and if you were the bottom third you were an air gunner.

A very kindly old gentleman told me that I did very well in the Link, and I said, "Oh, I don't know if it's going to do me much good; I understand they're going to send a bunch of us from our course as navigators." And he said, "Well, you'd like to fly would you?" I indicated how enthusiastic I was, and he said, "Well, based on your Link, I'm going to put you down," and I saw him write it down: "Highly recommended for pilot". But I went out as a navigator.

The Elementary Flying Training Schools

Graduation from Initial Training School did not always end in disaster for aircrew trainees with pilot aspirations. There were some disappointments depending on the progress of the war and the demands of the various units looking to the Plan for new or replacement staff, but on the whole, if a man had applied to be a pilot and if he had qualified through his courses at ITS, he would continue with the necessary instruction. He was then posted to one of the many Elementary Flying Training Schools located all across the country.

Student Pilot, RCAF: It was called "elementary" because it was the first of our actual flying training. We were in the Air Force, but the training was done by a club that had been reorganized to an Elementary Flying Training School, and this school was run by a civilian company.

Supervisor, civilian: They had a president of the company, and he had a secretary, and there was a Chief Flying Instructor who was a civilian and he hired his own staff. When the war came along, the Air Force advertised for pilots, gave them a check ride, and if they were good enough, one of the EFTSs would hire them.

Where I was, the CFI never *was* in the Air Force—he was too old—and he was paid as a civilian. He owned a fish packing company and he had learned to fly so that he could bring fish from northern Saskatchewan to his fish plant. The other instructors were Air Force pilots on leave without pay, and they had reverted to civilian clothes to man these Elementary Schools—and they were paid by the *school*.

There was one Air Force Supervisory Officer there in charge of the student flying program and he had an assistant called a Testing Officer. After so many hours they tested the students, then they would advance some more, and then they'd test them again. If the student was borderline and the Testing Officer didn't want to take the responsibility of failing him, the Chief Supervisory Officer would then give him a test ride and make the decision.

On these stations all of the students were Air Force; they had their own barracks and they were under Air Force supervision. The aeroplanes all belonged to the RCAF.

Instructor: We were in civvies and we were in the service—you bet we were, the RCAF—on loan to the EFTS. We had a uniform that resembled a cross between an airline pilot and a naval officer; they were really quite smart-looking, and we were often mistaken for an Air Force from a foreign country. We had some old World War One pilots as instructors. They were all exactly 39 years old because that was the cut-off age, but some of them were well into their fifties! I know they were, and a lot of them hadn't done any flying between World War One and then; they were old fighter pilots.

I think we were making $350 a month, which seemed to be reasonably good pay at that time.

Officer, RCAF: In 1940 I was an Assistant Chief Flying Instructor and we used to check out pilots for instrument flying and various other things to determine if they could instruct. I'd fly them around to see if they were competent to handle the aircraft and get out of spins and whatever. There were a lot of Americans: people who had graduated from the Boeing School of Aeronautics in California and people from Texas. Gees, we had them in the Plan from all over the States; I don't know where they came from. God only knows!

Student Pilot, RCAF: We went to No. 13 EFTS, which was St. Eugene, and it was run by the Ottawa Flying Club as a business. There was a chap who was the manager and I had an American as my instructor. He was a bit of a wild fellow who liked to fly low, and he scared the dickens out of me many times. He had learned to fly in the States somewhere, and he had come up to Canada, and they had just hired him on.

Student Pilot, RAF: Our EFTS was a British station, but it was civilian-operated. The maintenance and all the operation of the aircraft was done by Canadian civilians. The cooks and the people in the

IT WAS A REAL MIXED BAG

I had ideas about flying back in 1929; everybody did in those days—the world was full of Lindbergh. My education didn't extend beyond Grade Seven, but I got a commercial licence and I spent some time flogging around trying to build up time barnstorming and working out of the north flying old prospectors around.

In 1937 I was doing ground chores for the Montreal Light Aeroplane Club, but I really wanted to be an instructor. The Air Force had a reserve squadron at St. Hubert, and that's where they used to run the tests, so I got my instructor's licence there by flying around and demonstrating my skills to an Air Force Squadron Leader by the name of De Niverville.

Early in '39, before things really started heating up, the Department of National Defence gave the Montreal Club a couple of aeroplanes, a Fleet and a Gipsy Moth, and the Air Force started sending us provisional pilot officers for ground school and flight training.

In 1940, when the BCATP was starting up, several members of the RCAF visited the Club; they were opening what they called No 4 EFTS at Windsor Mills and they came to the Club to line things up for us to operate it. So the Montreal Light Aeroplane Club just sort of ceased operations and the chief flying instructor became the CFI at No 4 EFTS, and I went there as a civilian instructor.

The BCATP people built hangars and supplied us with about sixty Fleet Finches. A lot of the other instructors had joined the RCAF as sergeant pilots; they had previous private flying experience and had volunteered for the Air Force. They were given instructor's courses and then released and sent to our EFTS. They weren't wearing Air Force uniforms; we all wore a sort of civilian outfit.

The people that ran No 4 were all civilians, but the CO of the station was in the Air Force. However, I always answered to the chief flying instructor, and *he* was a civilian. It was a real mixed bag!

canteen and the people on the flight line were all civilians; the man behind the bar was a civilian and the girls who waited on the tables were civilians. But the instructors and the students were all Royal Air Force.

Student Pilot, RAF: Neepawa was a brand new station still under construction, and one of my earlier chores there was helping to build Tiger Moths. They had a whole bunch in crates right from the factory. The station had been "manned to establishment", but establishment was not supposed to be in the business of assembling aircraft, so they coerced the pupils into helping put these things together. All sorts of technical fellows fluttering around the edges, sort of thing: "Don't do this; don't do that," and "Ah—you've got it right, Sonny." They wound up flyable!

Student Pilot, RCAF: One of the interesting things about the civilian EFTS was the food. The first thing that they told you when you arrived on the station was to be careful how much you ate, since the food was so good. It was like home cooking really, far different from Manning Depot—it was really nice. We were out on the bald prairie and the flatness of everything was impressive.

Student Pilot, RAF: The Sergeants' Mess was still in the process of construction; people were hammering all night long—the most important building on the station. When they did eventually get the last plank in place, along came a huge truck, full of beer, and they needed six volunteers—you, you, you, you, you, and you—to help unload the beer into the Sergeants' Mess. So a little squad of us were deployed, grumbling, and we unloaded an *unbelievable* quantity of Dawes Black Horse Ale. And the old Flight Sergeant was a real slave driver: "Get the stuff in here, fellows!" Eventually we got the last case in and sank down exhausted, and he said, "Well, come on now—we'll open up a couple of these." About four hours later six pupils wound their way back to their billets. That Dawes Black Horse was really, really good beer!

Student Pilot, RCAF: When I got to the airport and finally saw those aeroplanes, I'm not sure whether I was excited or scared! But anyway, I knew then, "Here we are. We've got this far and we

When we first arrived at Windsor Mills, it was really exciting to see the line of yellow aeroplanes. There were thirty or so Fleet Finches that we were going to take our training on. It was a biplane made of canvas; it seemed to be canvas and bamboo.

When I looked at those Fleets I just loved them right away. They looked like something that would fly! I liked their outline with the radial engine and so on. I had a real thrill; I felt, "That's something I can handle."

are now going to start flying the Tiger Moth." The aircraft were for us to fly . . . and here we were. So fly we did, and had a lot of fun.

Student Pilot, RAF: The actual training process was half the day in ground school and half the day flying. If you were in ground school in the morning you might have three or four periods on different subjects, just like a school classroom. Then in the afternoon you'd go down to Flights. Each instructor had so many pupils and he would divide his time up between them according to a fairly mixed schedule.

Student Pilot, RAF: We went to No. 32 EFTS Bowden. Great excitement—see what they've got to fly—great, big Stearmans. It looked so big compared to the Tigers that we'd seen—great, big radial engine on it, 350 horsepower or something. And you talked; you'd get in there as a new class and the guys that had been there a week or two began to lord it over you, and tell you all about flying, and, "It's always terrible; they'll wash you out if you do this or that." Everybody's ready to tell you the worst of everything.

Student Pilot, RCAF: It feels pretty wonderful: "Gees, I'm gonna be flying. I'm gonna be an ace." They'd show us all the different procedures and how to start, before you're introduced to your instructor to fly. It was a wonderful thing; it was a *positive* thing.

Ground Staff, civilian: In the summertime we started as soon as daylight arrived, at four o'clock in June. Prince Albert is fairly well north and it was beautiful smooth air then, and they'd fly till the air got rough and then they'd close the flying down. The students loved it outside of having to get up. The instructors didn't like it too well, though, being out the night before. Then about four o'clock in the afternoon they'd start up again and they'd fly till dark. In the winter they flew daylight hours—there was no problem with rough air.

My Elementary was the Northern Saskatchewan Flying Club or some such outfit, and it was operated by a company for the Air Force. Our food was beautiful. We sat down six to a table in an Airmen's Mess—cereal, juices, everything; it was tops. Our billets were brand new, but the weather was colder than billie be damned and it snowed practically every night.

Student Pilot, RCAF: It was just a field; there were no runways, so there was nothing to get your direction on except the wind sock, and you tried to land into wind as much as possible. But you had to look out for other aircraft, because there were a number of aeroplanes in the circuit at the same time.

Student Pilot, RCAF: You never were delayed landing. If you were coming in on final and somebody was ahead of you then you just landed alongside him. There was never any problem with traffic; everybody could be flying and doing their own thing, and you just pulled off a little bit to land alongside, or take off. If somebody was sitting out there you just took off anyway.

Ground Staff, civilian: I've seen kids come in that got caught solo in a wind. They'd come in doing everything according to the book, and just before they sat down, the aircraft would start to go backwards. We had a little Chevy half-ton with a couple of guys in it and they'd go right alongside. As soon as that aircraft touched the ground they'd pile out and each man would grab a wing.

Officer, RCAF: They couldn't fly Tiger Moths at Lethbridge. Some of those days the wind was fifty or sixty miles an hour and your stalling speed on a Tiger was about 35. They could touch the wheels down but they had to keep the power on to stay in one place; in strong winds you could damn near come in like a helicopter. Lots of times they had fellows on each wing tip trying to keep the Tiger on the ground.

Student Pilot, RCAF: At Goderich you'd get some magnificent winds, and what the instructors used to do for laughs was to fly a square circuit. They'd get a bunch of us guys out holding the wing tips of a Moth and they'd take off, full throttle, and climb up, and then fly *backwards* across the field, close the throttle to descend, and then pour on the power to come in. And we'd all be out there to grab them as they touched down.

Student Pilot, RCAF: I was literally terrified of heights, so I wondered what I was going to do when I got into an aeroplane. I found it wasn't the same, except I couldn't stand stalls—I'd go, "Ooooowhooo," and do absolutely nothing. But I did like spins, you just sat there and the world revolved around you, however my instructor didn't like the way I recovered. I'd just give full throttle, and we'd be pushed way down in our seats. He told me I was going to kill him.

Student Pilot, RCAF: The first day my instructor took me up he said, "Well, we'll just fly around a little bit and let ya feel what it's like." And I thought, "Oh, gees, this is really nice," although I was kinda wondering a little bit about things. And then he said, "Well, we'll do a spin," and right when he said that I thought to myself, "Whoever told me I wanted to be a pilot?"

Student Pilot, RCAF: The very first thing they do is take you up and do a spin. You don't know what the hell he's gonna do and he says, "Now here, just

In the classroom we learned the theory of flight and other things about being a pilot that you have to learn on the ground. We did a lot of "hangar flying". You would sit in the Cornell and go through all the pre-flight checks. We had to remember Tom Mix Fucks Four South Carolina Girls. Tom was throttle; Mix was mixture; Fucks was flaps; and so on. One of the things we had to do was move the throttle all the way open and back again. When I finally got into the aircraft I had a twenty-year-old kid as my instructor, and when it came to "throttle" I did just what I'd been doing in the hangar—I pushed it up all the way. That's just like putting your foot to the floor in a car, and the aircraft went up on its nose!

Instructor, civilian: Any new courses you got, the first eight weeks was a real struggle, because you had four students that you were responsible for and none of them could fly on their own. They all had to learn from scratch, and you were the one that had to teach them. So you put in long days; about an hour and a half on each student every day. You flew straight and level; you taught them to do rate one turns; you taught them to climb and to glide, and to get the feel and understand the characteristics of the aircraft. Once they had that, in possibly two or three hours, then you would put them on take-offs and landings. You would have all your students at the same place at the same time, and you would do this for about three weeks—take-offs and landings, take-offs and landings, take-offs and landings— learning to land that plane. And then they had to be assessed.

pull back on the stick . . . back . . . back," and you know darned well something's wrong. "Just see what happens. Pull it back . . . back . . ." And you think, "Cripe, I wish you'd told me ahead of time what we were gonna do!" And of course you go into a spin, and what was it? Opposite rudder, nose down, and it came out so easily. It always looked so interesting when you'd see it from the ground. It was great.

Student Pilot, RCAF: You had to spin and you had to do aerobatics. We did rolls, loops, and stalls, and they put a lot of pressure on spins because you could get yourself into trouble with them. It was very hectic; it was a morning-to-night thing; it was a grind. There was not too much time for fun, because you had so many hours to put in and you had to accomplish so many things in that time. And then we also had the ground work: navigation; weather; this sort of thing. I can remember going to bed sunburned and tired, just beat!

Ground Staff, civilian: We looked after the Flight lines, bringing them into line after they'd been landed and taxied up to the tarmac—help the pilot park them. In those days you had to start the aircraft by hand, and I was mainly a prop swinger, although I did do a lot of maintenance under supervision, and all the aircraft had to be in and lined up and fuelled before they were stored in the hangar at the end of the day's operations.

The prop usually stopped straight up and down, and in this position it did not allow us to store the aircraft, because they were biplanes and the prop would not go between the mainplanes of another aircraft, so we turned them crossways. This aircraft had just pulled into line, and I hadn't noticed it and didn't know it was hot. I was just going down the line pulling them crossways, and this one hit and caught me across the hand. I swore it hit me fifty times before I got outta there, but I guess it only hit me once. I thought the hand was gone. I couldn't feel a thing—it was just numb—and I was afraid to pull the glove off for fear I was gonna take the hand right off with it!

Student Pilot, RCAF: Mr. Birch, my instructor, was a prince as far as I was concerned. He took you to heart and he was always interested in you. Some of the instructors didn't have a clue how a chap was doing at ground school, but Mr. Birch took your ground training almost as seriously as your flight training. He was always yapping at you if you didn't get a good mark in navigation or aero engines. Most of the chaps that I was in the barracks with, their instructors didn't care how they were doing in ground school just as long as they passed the flying, but Birch said, "You have to have both."

Instructor, RAF: I had four pupils, and you get very engrossed in them. I'd fly each one of them an hour and a half a day. That doesn't sound like much in a working day, but that's a lot of concentrated flying. You're doing seven or eight circuits with each one, so that's a lot of landings you're doing to get them off solo. And they were all different.

Student Pilot, RCAF: My instructor took great pains to explain something before we went up, which was a good way of going about it because you got the idea from him where he could actually use his hands to describe things. And before getting in the air he could show you on the controls what he meant.

There was a lot of evening work when you got down to your books. And if you wanted to do a few more charts or fiddle with required apparatus, you'd go down to the rooms that were equipped with that sort of stuff and you'd brush up on it. But by and large you did your studying in your room, sitting on the edge of your bed with the light on.

Student Pilot, RCAF: I had a US Flight Sergeant as an instructor. He was RCAF, but he had "USA" on his shoulder. He was fed up with instructing and he had little interest in helping you understand.

Student Pilot, RCAF: Air Force training was a numbers game. They drilled you in precision. You climbed at 65 miles per hour; you didn't climb at 64 or 66—it was 65. And if you allowed your airspeed to vary you were told about it. They had these numbers drilled into you so that you did your climbing turn precisely at 500 feet, not 550 or 490. You were so busy keeping track of the numbers you didn't have time to worry.

Instructor, RCAF: Discipline was such that you practically had to climb over into the front seat to tell that fellow to keep his speed at 70 miles an hour, because 69 wasn't good enough. And I think you turned out a good pilot.

Student Pilot, RCAF: All the instructors in the dark blue uniforms, the civilians, were darn good to you. They didn't quite have the father image—more like an older brother. They really wanted you to get along. But when you went on the twenty-hour check with the Air Force guy he was pretty severe. None of this camaraderie stuff; he was all "Sir" and that. I remember he said, "OK, we'll do a forced approach on this field." I thought I was doing all right, but he said, "You'd better try a sideslip." So I sideslipped, and he said, "Not like that! Like this!" and, Jesus, he stood it right on its wing tip and down we came. And he said, "All right, you pull it out." So I pulled it out. But he had an inhibiting effect on you with all this rank. I didn't like it, and I was as much afraid of him as I was of what could happen in a crash.

Student Pilot, RCAF: Most of the instructors were a bit older than the average student, but they treated us like gentlemen. They weren't the disciplinarians that we met on the parade ground—it was different in the air. The one that I had was quite an efficient fellow, and very effective. He always pulled little stunts on you. If you were coming out of a loop or something he'd tell you to force land. I didn't get the impression that they were too strict or too tough—they were good.

Student Pilot, RCAF: Mine was very demanding, and I remember him demonstrating very clearly the effect of cross controls in a steep turn. In the one case you had top rudder with bottom stick, and the instant stall and the flick to the outside. And then the reverse with bottom rudder and top stick, where you violently spun in. His demonstrations gave me a lot of respect for that ball in the centre.

Student Pilot, RCAF: My instructor was a nice guy, but he'd been instructing too long. As we taxied out we'd have a nice little chat on the intercom, but then as soon as we opened the throttle for takeoff, he'd become a mad man. In the air he would curse you, and he would snatch the controls

I SAY, OLD MAN

Instructor, RCAF: We had Royal Air Force and RCAF instructors, and it was an interesting comparison. In those days the language in an aircraft when you were teaching somebody to fly wasn't the best. We were swearing and cussing—"You stupid so and so . . ."—and whatever, to get our students to learn. We used to get our guys solo in about eight hours, but our English instructor friends came over and they had a more gentlemanly way about them than most Canadians. If the student wasn't keeping his airspeed right, they would say, "I say, old man, now can you do a bit better than that?" Their students were soloing in about ten hours, and a lot of the difference was in the system. We tried to make people sharp and on the ball right from the word go, and they did too, but in a nicer way.

out of your hand . . . so much so that at one time I went to my Flight Commander and said, "Either give me a change or wash me out." I had about 25 hours with this man when I felt that I couldn't go on with him; we'd go up for an hour at a time, and by the time you'd finished an hour with a chap like that you were in pretty bad shape.

Student Pilot, RCAF: The first thing my instructor said to me was, "Do you want to go low flying?" And I said, "You're damn right!" He said, "Well, don't ever go low flying unless you tell me. Whenever you get the urge you tell me, and we'll go low flying. That'll keep you out of trouble." I said, "I want to go now." And that's what we did for 75 hours—we low flew!

Instructor, RCAF: The objective of an instructor is to generate a pilot, but there is no point in generating one who is going to be dead within two weeks of getting his wings because he was stupid or stuck his neck out. I learned from my instructors you don't fool around. You learn to keep your aircraft in one piece and it keeps you in one piece.

Student Pilot, RCAF: These instructors were all very fine men—gruff, but nice fellows. Many of them had been flying a lot in the bush, and I guess to fly a Tiger Moth was pretty mundane to them, but to us it was pretty thrilling—screaming over those little tubes.

Student Pilot, RNAF: We had a blow horn. You could speak into a funnel—you have it in front of you and it's hung up by a spring—and you talk into it by putting it up to your mouth. It was kind of difficult at times, particularly if someone had been sick in the aircraft prior to your trip—the stench was unbelievable and you had to open the coupe top about a quarter of an inch so you could get some fresh air. Then it was almost impossible to hear what the instructor was saying because of the noise.

Instructor, RCAF: Everybody had their rubber mouthpiece which they stuck in their belt. It fitted over a metal pipe right in front of you, and you yelled through it. And right in your helmet you had an earpiece that was attached to the student's pipe. It was not electrical in any way, and if it wasn't too noisy, you heard.

Student Pilot, RCAF: It led to frustrations, because with canopies and all sorts of things vibrating it was difficult to hear, and the instructor would certainly get angry if you didn't respond.

I felt that I was training for the First War when they threw bricks out the back of those small SE 5s. You'd get up as high as you could and then turn off the engine, and you'd just float like a bird. Sometimes the mechanics had to come out and bring us down because the wind was too high and it wouldn't land, and they'd have to force it down. I was thrilled to be part of it.

Instructor, RAF: In winter they used to freeze up. You'd breathe into these things and they'd get an ice clog. When I did my instructor's test I got this rather nice fellow and he put me through the usual routines, and I was yelling my head off to him. After we got on the deck he said, "OK, that's pretty good. I couldn't hear what you were talking about half the time, but you seemed to know what you were doing."

Student Pilot, RCAF: You have a real *bond* with these instructors. I always thought of mine as a real friend, because he was the first person to actually take me out to that aeroplane and sit me in it and say, "All right, we're gonna go," and he finally got me solo. I figured I was getting close, but I suspect what he thought was, "This guy, we gotta make him or break him, so we'll let him go." Because when he looked at me I think he was sort of saying, "Good luck. Oh, gees, I hope you make it."

Student Pilot, RAF: I taxied 'round to the dispersal and I started to take off my straps. My instructor said, "No, no. Keep your straps on; I want you to go solo." I thought, "Oh, God—isn't that great!" And he gave me that, "Now, if you don't think you're gonna get down, I don't care if you go 'round ten times. If you're landing long, that's no good." It didn't worry me. I knew I could do it, and I remember getting off, happy as a lark. I knew I had no problems with it. Of course, the place was buzzing with aircraft; there were always ten or twelve on that circuit. But it was a new world: nobody behind me; nobody to shout at me. I had a feeling of accomplishment; even though I hadn't landed it I knew I was away.

Student Pilot, RCAF: It was a very strange feeling when I looked and there was nobody in the back seat—nobody there; you're on your own. That first trip was really something. You climbed to 500 feet and did your rate one turn to the left and got to 1,000 feet and turned downwind and then crosswind and then landed. I was so busy getting everything sorted out, that I was at 1,000 feet before I remembered to do my rate one turn to the left. I climbed straight on out to 1,000 feet, and afterwards my instructor said I had given him a few anxious moments.

Student Pilot, RCAF: I had a very normal check flight. We came in and landed and he said, "OK, taxi over to the hangars; I'm getting out. Take it away and good luck." Fine. Hell, *he* didn't have a care in the world. But I took off—scared, apprehensive—gee, you're up there all by yourself in a Tiger Moth. And I did a circuit, prayed that there'd be an open place for me to land, and I didn't have any problems. Fine—I had soloed!

Student Pilot, RCAF: I didn't think I was ready and I wasn't all that cocksure of myself. But we had done some circuits and bumps and we came in, and he said, "OK, away you go!" I said, "Who, me?" . . . "Yes, away you go." So I did. I just wanted to get it over with and get back to the hangar and celebrate a little bit. A number of us were soloing about the same time, so we all went into Goderich and had a couple of beers, and thought we owned the world!

Student Pilot, RCAF: He said, "I want you to go. Just take off in your turn, make one circuit, and land." I knew what he was saying, but I thought, "I'm not quite ready." You know what I mean: "Let's talk this over with a cup of coffee." Then I thought, "Well, he knows, and I'm damn sure I can do it." You just feel like singing; you just feel, "Here I am, by myself," and you know then that if you make a mistake there's nobody to help—*you're going to bring this aeroplane down.*

Then of course they threw you in the shower—it was part and parcel of the whole thing that this great Air Force should throw you in the shower with your clothes on. Then you'd all buy a drink; you felt like *really big stuff!*

Student Pilot, RCAF: That evening back in the barracks all us solo students had a cold shower—but they always waited until you got all dressed. I had a date this night with a Miss Elkin from Three Rivers, and gosh, I was going to tell her I'd soloed. I thought I was getting by, but no, I wasn't. They'd wait until you got all dressed up, then . . . in you'd go. I got the cold shower and then had to change my uniform.

Whatever the students might have felt about their latest achievement, their instructors had thoughts of their own.

Instructor, RAF: There's a little trepidation. You're relying on your own judgment and you know in your own mind that this kid's a bit weak in this way and that, but on the other hand, "Is he safe?" Once you've satisfied yourself of that then he's on his own. One worries a little bit, but as long as you're confident that the pupil has adequate control of the aircraft and the ability to regain control should he lose a little bit—like swinging on landing and keeping a lookout for other aircraft and this kind of basic stuff—once you know he can do that, then the refinements come later.

Instructor, RCAF: Finally you got him to that point where you'd been yourself and you decide, "Well, I think he's ready." And usually at that stage, depending on your experience, somebody else would fly with them—one of the assistants or the Chief Flying Instructor—to see if he was really ready.

Instructor, RAF: I was a new instructor, and they have to assess your capabilities. I think if I'd been the Flight Commander I'd have done the same thing. He said, "I'll give him a check ride." And I watched him. I thought, "Gees, I hope my guy does it all right!" He did one circuit and taxied in, and I saw the Flight Commander stand on the wing, pull the coupe top back, and speak to this boy, and I knew he was going to send him solo. He took his parachute and jumped off and came in to me and said, "He's gonna be all right." So I went out and sat on the edge of the apron . . . and watched.

Instructor, RCAF: *You were nervous, but you felt confident that the kid could do it. You knew they were bright and competent and had what it took, but you still worried with a lot of them: "Boy, if he cranks that thing up, we're all in trouble." But after you'd instructed for a while you knew they were OK, and you'd just step out and say, "Away you go." And when he would return, like my own instructor, you're there to shake hands, and everybody congratulated everybody else. And that's very important, and encouraging, too.*

You'd do two or three a day like that sometimes. There's a great bunch of them with a new course—fifty kids—and every day there was always somebody going solo.

Student Pilot, RCAF: The instructor turns you over to the Examining Officer, and he takes you up and asks you to do certain things. He is the one that decides whether or not you are ready to go solo. In his judgment I was ready to go, and I went.

Instructor, RCAF: One fellow was one of these people who can hang by their ears from high bars—beautifully co-ordinated—but in an aeroplane he was a dead loss. He was ham-fisted. You'd say, "Let's have a little left bank," and to him that meant, "Push it to the wall with all your strength!" But he soloed. I stood there and watched him all around the circuit and didn't take my eyes off him. I remember keeping my fingers crossed and saying three Hail Marys.

Instructor, RCAF: I always felt very responsible about my students. I was a tough disciplinarian: "There's a right way to do it and a wrong way to do it, and you are going to do it *right*, or you are not going to do it at all." That is what I had learned from my instructors, and I thought it was good training. Absolutely rigid on the discipline.

Student Pilot, RCAF: I was beginning to get concerned, thinking, "I may never go solo," because my instructor was never inclined to send a chap off if he wasn't fully satisfied. But the day came when I seemed to be doing landings and other things right. And the procedure on our field was no discussion or anything—the instructor would simply get out, and in his hand he had a long red streamer which he attached to the port interplane strut. I think this meant, "For crissake get out of the way—this guy's solo!" But it was a great thrill to see him tie that streamer on your strut, and he walked back to the cockpit and he said, "Burford, don't bend it; it's my favourite aeroplane!"

Ground Staff, civilian: I can't remember where the pennant went. I know it was red, and some students would forget to take it with them when they were flying solo—or the instructor would forget to take it when he was going to turn them solo—and we'd have to run to the Flight Office and get them one. They scrubbed it after a while.

Student Pilot, RAF: After my solo I had a photograph taken, and I look like I had just won a lottery. It was a strange emotional mixture: of great joy—at last I'm going to do it; of great fear—because you are not sure. There had always before been someone up front, and if something went wrong he could look after it. Confidence and yet apprehension. But I put it down, and I was able to go back and tell everybody that I had done it. I was trained!

*****

Ground Staff, civilian: They didn't have snow removal equipment. They just smoothed the snow and rolled it—just as hard as concrete. They worked right through every snowstorm, and sometimes they were kept pretty busy.

Flight Commander, RCAF: Getting on skis was different, and it was sure hard to taxi. We weren't on runways, just a field, and all we had to show direction was a half-ton truck facing into the wind. And you kept moving it back and forth as the wind changed. The kids could land pretty well anywhere on the field and then taxi around the outside.

Ground Staff, civilian: When the students first went solo flying on skis it was a little hazardous. They didn't have any brakes and they used to get off in the deep snow; they'd sink down, and we'd either have to go and get a man on each wing, or if it was *really* off you'd have to get a tractor to pull them up. And we had to stay on the wings to guide them to their parking spots.

Ground Staff, civilian: In thirty-below weather we couldn't turn the engines off. An instructor would come in with a student, and there'd be another instructor and student all ready and waiting to get in. If you shut the engine down for any time at all it wouldn't start. Of course, we had to turn them off to gas them, but we did it very fast. That oil used to get pretty thick in a big hurry.

Student Pilot, RCAF: Every morning a farmer was out with horses and a manure spreader, putting manure on the field so that you could see the black against the white snow. Sometimes they had two or three teams going, and whichever way the wind was they would put out strips. Manure was the handiest thing—they couldn't be digging dirt in the middle of winter.

We flew any aircraft that was assigned to us. Some had wheels; some had skis. Sometimes when the surface was hard and icy, when you set that

Ground Staff, civilian: There were no electric runway lights, so we used goose flares fired with oil. They were like a tea kettle with a great long neck on them and filled with kerosene and a wick, and they were the filthiest things that ever happened; they were just covered with carbon. You ascertained the wind direction from the sock and laid the flares out in a T formation, and you flew into the T. I think we laid ten flares on the main leg and four across the top.

aircraft down with skis on you couldn't describe the noise. When you touched down there was a tremendous amount of clatter.

Student Pilot, RCAF: At Malton we were using skis in the big triangle inside the runways, and Malton was the Toronto base for Trans-Canada Air Lines, so we had lots of red flares! One chap overshot and actually hit the runway. He flipped on his back when his skis reached the concrete, and the engine was stuck in deep snow. He told us later that the hot engine was hissing, and he thought he was on fire, so he just pulled his straps and came down about three feet onto his head on the concrete. When they got to him he was out cold, but they bundled him into the ambulance and took him to the hospital. Within an hour he was back and into another aircraft, and away.

Student Pilot, RNAF: I was up on skis on my own doing some aerobatics. A ski flipped up in front of a wing, and that sort of threw me a bit. A cable that holds it in place had broken. I had to use quite a bit of rudder to keep the aeroplane going straight, but I came back to the airport and flew around to notify everybody that there was something wrong. They turned out the ambulance and the fire truck and everything was lined up as I was going in for a landing. As I came in and the speed dropped, the drag was reduced on the ski and it came down a little bit, so when I landed on the one good ski the other one just flopped down, and I stayed going OK.

*****

Ground Staff, civilian: During the latter part of the training plan there wasn't the pressure to get the fliers into Service Schools and on to active duty overseas. They were catching up, and so they decided that they would give the elementary students ten hours of night flying. This was really to be circuit flying only, and dual.

We had eight aircraft designated for night flying and they were specially fitted with navigation lights and a signal light at the bottom, a telegraph key, and a twelve-volt battery for power.

Student Pilot, RCAF: We had a single flare path in the grass at Prince Albert—just little flares put out in a row—and you tried to keep that on your left. We didn't get much night flying, just enough to whet your appetite on it, and we never did any solo at night. I guess they didn't want to lose any more Tigers than necessary.

Instructor, RAF: We had no radio, but the Air Control Pilot (ACP) at the end of the runway had an Aldis lamp and he would signal. You came downwind and you signalled your letter—mine was "P"—with your downward identification light, and the fellow at the end of the runway would flash a *green* "P" back at you. It meant, "OK, you're the next to land." There were six of us on the circuit at night, and he knew where we were. There'd be one landing, one taking off, three taxiing, and one on the downwind leg. The ACP had to know where they were, and the instructor would say to the pupil, "OK, flash the letter, boy." The ACP would flash back, and you'd say, "Did you get a green?" . . . "Yes, Sir, got a green." . . . "OK, down you go then."

Instructor, RAF: Assiniboia sent students solo at night. These students never thought they would solo, but we got them when they were well advanced in their regular course, and they caught on pretty quickly. They'd done about 35 hours and they had good airmanship qualities and they could *handle* the aircraft. But night flying—judging the flares, your height, your speed, and your drift was all different.

Ground Staff, civilian: We had flying instructors who didn't like these circuits and bumps in the dark, so they'd take off cross-country. All at once instead of having six aircraft in the circuit, you'd only have four. And you looked, and you looked, and gees, you can't find them. Then pretty soon there's five on the circuit again, and then there's six back. And there's no way you could get those instructors to admit they went out fooling around someplace.

Student Pilot, RCAF: I remember my first solo night flying. As long as I knew my instructor was in the back it was fine, but finally he said, "OK, take off, away you go." So I took off in this darned Cornell, but all of a sudden it got very, very dark, and I thought, "Oh, boy, I'd better get down quick." I was on my second try at a landing and I levelled off too high . . . and went around again.

I got it down at 3:30 the next morning. I had remustered into aircrew and my aero engine experience helped, because I pulled up to about 7,000 feet, leaned the mixture, and thought, "To Hell with you guys. I'm gonna stay up till I can see the runway in the daylight."

Ground Staff, civilian: I saw an aircraft come in with a solo student and I knew he'd landed hard; I *heard* him hit! So I jumped in the truck and tore out, but before I got there he was gone again. When he came 'round the next time he said, "I think I had a hard landing. Would you check the aircraft?" And honest to God, you could move those wings up and down six inches . . . and he'd made a circuit that way. I thought, "It would take the wings twenty seconds to catch up to the fuselage!"

Instructor, RAF: One night, it was in the Fall, I was in the back seat. The pupil was doing all right, but all of a sudden I saw him duck his head into the cockpit and there was an awful noise and a bang! The engine started to run rough. I thought, "God! We've flown into the back of somebody;" it was always a haunting fear. I grabbed the control and shouted to the pupil, "What's happened!" He said, "We've run into something!" I thought, "Oh, Jesus . . ." From the glow of the nav lights I could see a whole sea of things. It was a flock of geese; we'd flown into these Canada geese. So I sideslipped it down and landed. We got out, and 'round each oleo leg was a goose! One had gone through the leading edge of the plywood wing—in as far as the gas tank. One had struck the prop and broken a bit off it, and part of its foot was in the brass fairing near the tip. What a bloody mess!

*****

Instructor, civilian: You'd have your very initial students—the juniors—and then the seniors. That was about the variety that we got. You'd get pretty fed up putting in six hours day after day and doing the same things over and over and over again. The part I enjoyed most was the aerobatics.

Student Pilot, RCAF: My first shot at solo aerobatics was a dismal failure. My instructor said, "Go off and try a few things." Now, my legs are rather short and I wasn't ready for the negative G; I didn't have things adjusted right. I tried a roll and got on my back, and my feet came right off the rudders and landed up underneath the instrument panel. They just seemed like a couple of hunks of lead, and for a moment I seemed to be hanging on the control stick—upside down. It seemed that the only thing between me and oblivion was hanging on to that stick, and I couldn't seem to get enough movement of anything to get it rolled around. The only thing I could do then was to half-loop out of it, and for the first few moments as I came around the half loop I forgot to close the throttle. Of course, I had had quite a bit of power on for the slow roll, and did that Tiger ever wind up! I pulled her out awfully gently and thought, "Just don't do anything violent."

Student Pilot, RCAF: One day my instructor said, "We've done spins, so go on up and try some on your own." So I'm up there by myself. I was at 5,000 feet over the Ottawa River and I went to do that first solo spin. I had the power back and the nose up and I was just about to go into it . . . then I shoved the throttle open and climbed up another thousand feet.

Student Pilot, RCAF: I had trouble with aerobatics. I'd try to do a roll off the top of a loop or whatever the exercise was, and I wasn't doing it the way it should be done. My instructor was a very kind person—I never heard him blow his top, never heard him scream or yell—and I would talk to him about it. And we'd go up and he'd say, "OK, now do it." I'd do a loop . . . fine. I'd do a roll off the top . . . fine. He'd say, "What's the matter?" I'd say, "It's not happening this way when I'm alone." He said, "You're lacking in confidence. You've proven to me that you can do it." And so he just talked me into doing it, and eventually I mastered aerobatics.

Student Pilot, RNAF: There were scattered clouds, and I hadn't had any instrument flying, but I found a hole and got on top of them. I was doing loops—one loop after another—and the Cornell doesn't hold altitude; you lose a little altitude every loop. On one of those loops I went into the

WHERE IS MY KID?

Instructor, civilian: They were real anxious to get pilots trained, and we had two courses going all the time at different stages. So you'd have a couple of your students out flying on their own doing solo practice, and you'd be giving dual to the others. We had a real keen type who used to issue flight authorization chits for these solo students, and one day after everybody had gone flying he saw a student left over in the flight area and an aeroplane that wasn't being used sitting out on the tarmac. He assumed that the student was on a solo course and that he was a straggler that had been off sick, so without asking any questions he wrote out an authorization. The kid actually had about an hour and a half dual, but he had enough guts to go out and get in the aeroplane and take it off. It wasn't long after he was in the air that his instructor turned up and . . . "Where's my kid?" . . . and here he was in the air with little more than an hour's dual, and everybody knew what shape he'd be in when he got back down. There were sure some anxious moments.

71

clouds and I blanked out completely. My first instinct was to pull up to get up on top. And I kept on pulling and pulling, and finally I got out of the cloud. But the sun was down there, and I was upside down on top of a new loop.

Student Pilot, RCAF: We were using seat pack 'chutes and we had training in their use to the extent of being told, " Don't abuse it, because it might be what keeps you from hitting the ground awful hard." But in terms of actually using it: "You'll find out the first time!" There was a very stiff fine for abusing them, and if you blew it accidentally, that cost you more.

The guy in charge of the parachute section was Air Force, but most of the people working in there were civilians, and the 'chutes were periodically turned in for inspection. Parachute silk was a good trade-off for almost anything, and I remember one that was opened contained a blanket.

Student Pilot, RCAF: We always had a paper bag handy. My instructor gave me strict orders that if I was going to be sick, "Don't put your head over the side. Put your head between your knees ... and into the bag!" We had one instructor that I'll remember forever. The student had slipped the coupe top back and put his head over the side, and the poor instructor in the back got the whole blast. When he stepped out of that aircraft he couldn't see out of his goggles. He was a mess from head to foot.

Student Pilot, RCAF: The penalty was you had to clean it up yourself, and having just lost your cookies you were in no shape to do that. The alternative was to buy one of the hangar line guys to do it for you—if you had enough money. But nobody ever had any money, so usually you just did it yourself.

Student Pilot, RCAF: One of our fellows who was a good flier was continually airsick and they were going to wash him out—no question about it. The Medical Officer had a talk with him and said, "OK, take these and I think they'll cure it." And sure enough, they did—he never got airsick again. One day I asked the MO, "What did you give him?" He said, "Nothing; there was absolutely nothing in the pills—everything was in his mind."

Instructor, RAF: You don't have too many instruments in a Tiger Moth, so our instrument work was strictly needle, ball, and airspeed. We had a standard system where the student sat in the back seat and you had a hood that pulled over so he couldn't see out. The instructor sat up front as a combination instructor and safety pilot, and the manoeuvres that you did were really elementary: straight and level; level turns; climbing and descending; and climbing and descending turns. There was nothing complicated.

Student Pilot, RCAF: Your instructor would hood you and then put you through some unusual manoeuvres to get your ear canals thoroughly mixed up. Mine would then let go the stick and say, "You have control!" He taught me, "Ignore your feelings and depend on that needle and ball." He even put me into spins under the hood, and that was just the greatest training ever—I just loved that.

The Stearman (centre) was a powerful thing for doing aerobatics. You could do lovely rolls with it, but it didn't have a shoulder harness, just a lap strap. It went across your knees, and if you didn't have it tight and went on your back, you'd feel yourself drop! It just had a big strap three or four inches wide almost across your crotch, and if you were doing aerobatics, you made damn sure that it was tight. I remember not having it tight and getting the fear of my life because I dropped an inch or two.

Student Pilot, RCAF: Those old Tigers had the coupe tops on them, and I remember once pulling the coupe top forward, but probably not securing it properly. We got to about 500 feet and made the climbing turn, and all at once, Bang!, and away went the coupe top. I remember the instructor grabbing the stick, completing the circuit at 500 feet, putting the aircraft down, taking it over to flights. And, God, he must have smoked ten cigarettes.

Student Pilot, RCAF: A directive came out that we would take off with coupe tops open. In the climb I held the stick between my knees; on the left you've got the coupe top release and then you've got the overhead U handle. I got hold of the handle, but it wouldn't unlock. It was getting a bit rough so I got the machine settled down and headed towards the practice area, and then I started to struggle with the canopy again. The second time, nothing worked. It's pretty hard to fly in rough air with your knees, but I got the machine settled again and I reached for the U handle, and before I even touched it there was a big *whooomph* and a bang. The whole canopy had torn off and smashed into the fin and rudder. And, oh, boy, suddenly you're in that rear cockpit without the canopy and you're sticking away up over the fuselage, and you feel like you can reach back and touch the tail feathers. I looked, and the canopy was going straight back in the slipstream, so I turned and followed it and watched it settle into the bush. Then I flew back.

I felt very embarrassed and silly, landing and taxiing up with the whole upper part of my body sticking up out of the fuselage, but I reported in to my instructor: "I'm sorry, Sir; I've lost my coupe top." Well, he played it real straight and made me sweat. He told me just how long it was going to take me to pay for that coupe top, and I said, "Well, Sir, I know where it is; I can show you!"

Student Pilot, RCAF: We had to do a dual cross-country, and my instructor was an absolute expert on navigation. He had done the run so often that he had figured out it took one and a half cigarettes to get to our destination, and if you had a head wind, two to get back.

Student Pilot, RCAF: I had a girlfriend who lived south about twenty miles, and I decided to go down and beat up her farm. The way I navigated in those days was to climb until I could see where I was trying to get to, and then I'd come down. I guess I let the engine cool off too much on the way down, but I just kept going, right at the barn, and as I was waving at my girlfriend the engine almost quit on me. It coughed three or four times and then caught, and I missed the barn and flew home, saddened by the poor show, but wiser.

Student Pilot, RAF: They never let you land a Stearman on the runway in case you ground looped. The instructors would do it, but not the pupils. But I got sent on my first solo cross-country down to Currie Barracks, which was No. 3, SFTS, RCAF, just outside of Calgary. My instructor said, "When you go to Currie, you'll have to land on the runway." If he had told me I was going to the death cell that night I couldn't have been more upset. I thought, "God, the stories I've heard about Stearmans ground looping and swinging off runways." All the way down to Calgary I thought, "How will I ever get this machine down?"

Then I could see the airfield. They had Cessna Cranes there and of course we had no radios. So I got into the circuit and I'm on the approach, and by God, they fire a red at me and a Cessna Crane comes creeping in underneath—that was the first thing that upset me. So round I go again in the Stearman and get lined up with that runway again: "God, here we go!" If I ground looped it I knew I was in trouble. But I got that thing down on the runway and I taxied in, booked in to the control tower, and all these guys off the Cranes came and examined that Stearman. It was a strange aeroplane to them, and I walked around it as if it had been a Spit XXI! "How do you like flying these?" . . . "Money for old rope!" Well, I go back to Bowden and tell the gang that I've landed on the runway, and, oh, you'd have thought I was Douglas Bader: "You really landed on the runway?" That was a great boost for my flying and I never looked back.

Student Pilot, RCAF: I'd been flying maybe twelve or fifteen hours—I hadn't been solo very long—on Fleet Finches with the Kinner engine. In hot weather you had to look out the sides because the cylinders in front of you had little tappets and all the oil came off those onto the windshield.

I had a gasket blow one day and I didn't have all the cylinders working, but with pumping the throttle the old thing would juice up a bit. I was limping along and losing height and I wanted to get back to the field, which was in view, but I was coming in against the circuit. I was high and turning in opposite to all the others, but I didn't have any power to do much about it—I had to land. Someone had told me about a slipping turn, so I tried one and lost a lot of altitude in a hurry, but I landed about two-thirds of the way down the field. I went off the edge and went over a few furrows and bumps, and the crash trucks roared out, but I didn't ground loop and I didn't break the prop. I had to report to the CFI and I told him there wasn't much I could do about getting back, and because of all the other aircraft I came in high and landed as far out in the field as I could. But to be honest, I didn't have a clue where I was going to land—I was just lucky to get down.

A year or so later I had a friend who was teaching airmanship in ground school, and part of his course was about a new recruit who had an engine failure one day, and who had "the presence of mind and the qualities of airmanship to come in and land long so as not to give problems to the other students in the circuit." I was part of history! If they had only known the facts.

*****

Officer, RCAF: In the first few months of '43 the Air Force took over a lot of the Elementary Flying Training Schools, put all those civilian instructors in Air Force uniforms, and sent a whole gang of people like myself to be Flight Commanders and show them how to run the Air Force.

Instructor, RCAF: I got to Uplands about the end of 1942 and was given a few students. In February they were disbanding the civilian-operated Elementary Schools and picking instructors and opening up Air Force schools for elementary training. There was a changeover right then because the ultimate goal of the Training Plan was that all personnel would be RCAF and that when we had enough instructors all the Elementary Schools would be operated by the Air Force.

I was sent to No. 11 EFTS, Cap de la Madeleine. There were civilian instructors there in the ground school, and the Chief Engineer that looked after the maintenance was a Quebec Airways man, and while I was there, civilians still maintained the aircraft.

Instructor, RAF: The RAF and the RCAF had different philosophies about instructing. In the RCAF, every few hours a pupil is tested and he *has* to have achieved a satisfactory level of proficiency, and if he hasn't, that's it! In the RAF they gave us a lot more leeway to tutor pupils along depending on the individual's ability to learn.

Student Pilot, RCAF: My instructor was a Pilot Officer in the RCAF. He had gone through the Training Plan and had been chosen as an EFTS instructor. This was something that everybody was afraid could happen to them. They'd say, "There's one thing I don't want—to be an instructor." This man was frustrated; he should have been a fighter pilot, and I think a lot of times he'd take out his frustrations on us. But I did learn to fly well under his direction. We'd go up and he'd say, "OK, how do you feel—your stomach's OK?" And then we'd drive that aeroplane all over the goddam sky.

Student Pilot, RCAF: At Malton we were using Trans-Canada Air Lines runways. That was the only operator that came in there, and you were looking in the circuit for Tiger Moths which were small, and the circuits were tight. You would see a spot way over and never think that he was on the final approach so you'd come on in. There'd be red flares all over the place, but some chaps would look around, couldn't see any danger, and they'd go on in and land anyway, and TCA would have to zoom over for another pass. You'd get asked, "Didn't you see the flare?" And you'd say, "Yes, but there was nobody near me so I thought it was for somebody else," because TCA would drag them in from such a long distance compared to what we were doing on a circuit.

Student Pilot, RCAF: I undershot the field one day. The terrain was rough, and when I touched down I slid along on the grass and then hit some ruts. I guess it had been ploughed some time before and she just tilted right up on her nose. I couldn't do very much about it except get out, take my 'chute, and hike across the field back to the hangar. Somebody said, "Where's your aircraft?" I said, "It's out there. I couldn't pull it in myself." I suppose that accounted for a few demerits.

Student Pilot, RCAF: The whole thing in this training for a pilot was the matter of washing out. You lived with that. It was a real worry that you had to pass every exam, you had to pass every flight check, or they were gonna wash you out.

Student Pilot, RCAF: It was something that was always uppermost in our minds. We actually didn't fool around too much with the aeroplanes; we were desperately aware of getting caught, and if you did, it was game over.

Student Pilot, RAF: A lot of people were washed out, maybe ten percent at EFTS, and it was pretty grim. We were all in those long huts and people would say, "Oh, gees, I'm being sent back to Trenton." The fear of getting washed out never left you; nobody was playing cheekie, because if you

THEY WERE BLOODY WELL RELIEVED

Testing Officer, RCAF: You couldn't take your hands off the stick of a Tiger Moth, any time. You had to *fly* the bloody thing *all* the time. If I found that people were sadly lacking in co-ordination, I'd just have to tell them, "Look, this thing, it's a lot greater than you or me. It's the Service, and it's Canada, and it's a lot of money that's at stake. And it's your life and whoever might be crewing with you." The majority of them accepted it. There were a few that were reluctant, but there were others that were bloody well relieved to have me tell them that I didn't think they could make it. Just looking to the future, they couldn't visualize themselves getting into a Harvard or an Anson or something and increasing their responsibility. They were still aircrew and they could go on as navigator or gunner or bombaimer.

just put one foot out of line they had you. Everybody had the feeling that they wouldn't stand for any nonsense, and if you started flying under bridges and such you just didn't have a leg to stand on. These machines had great numbers on them, and farmers would phone in and say, "Some son of a bitch with this number . . ."

Student Pilot, RCAF: They allowed ten hours for solo. If the odd person was having difficulty the instructor on his own initiative might give him a couple of hours more, but generally they wouldn't waste the time. If you hadn't soloed by ten, most people just washed out, and that was the end of your pilot career.

Student Pilot, RCAF: The most constant worry of an airman-to-be was that somewhere along the line the instructor might say, "Sorry, it's not working out. You just don't have the ability," and you'd be washed out. Perhaps you were with the same fellow from Manning Pool, to guard duty, to ITS, to Elementary, and within ten days or two weeks he's gone—just couldn't do it. The instructor had to make the decision that he could or he couldn't. That was it!

Student Pilot, RCAF: One chap that was washed out, he thought he was an out and out failure—just the fact that he had to leave our group. You're like a hockey team where if one guy broke his leg you felt as sorry for the team as you did for him. We were one group, and we were all going to graduate together, and we were all going to win the war together.

Student Pilot, RCAF: I never knew whether I was six feet up or 600, and instructors don't like it when you level off for a landing at 600 feet. One side of the field had a fence, and I made the discovery that once I got over the fence, if I pulled the stick back I settled down very gently.

I was pretty near ready for solo, and on my second time 'round my fence was coming up towards me—I just loved that fence—and I just got over the top of it, pulled the stick back, and made a lovely landing. We go 'round again and there's my fence. I pulled the stick back and did it again. That's three times—he's going to let me go solo . . . I've made it . . . I'm a pilot! What a wonderful feeling. One more time around. We're coming in and my fence is coming up towards me, and all of a sudden in my ear on the intercom the instructor yells, "Stay away from that God-damned fence! You're no friggin' crow!" It ruined my whole day, and that was it—I got washed out. As we walked back to the hangar he said, "I'm afraid we can't give you any more time." . . . just like that. I cried; I just wandered around in the hangar amongst the aircraft. I couldn't talk to anybody.

Instructor, RCAF: As an instructor it wasn't my job to actually wash out a student. I would recommend to the Flight Commander that I didn't think this fellow was going to make it. He would check ride him, and if he agreed with me the lad would be "cease flying". I only had one, and I felt that I was going to save his life because I thought he would kill himself.

Student Pilot, RCAF: I remember one that departed—a young American fellow. He was a smart-ass; he took off in the back seat and landed in the front seat, and they took a dim view of that. So they sent him back to the US of A.

Student Pilot, RCAF: One time I landed to have a leak. I landed in a field and somebody got my number—I just landed. I thought, "Well, gees, I can land." You're not supposed to land until you come back to the airport, but I landed and had a leak.

Then there was a spin. I left a spin in too long when the Chief Instructor took me up for a test, and I did some other crazy things. You're supposed to look both ways for other aircraft; I didn't look and I looped around an instructor with another student pilot. I couldn't understand when they said, "Why in the Hell did you do that loop? It's a basic to look around for other aircraft. We can't have that going on—you're endangering lives." And they said, "I've noticed you landed; you knew that was wrong." And I said, "Yes, Sir, but I needed a leak." Then the Chief Flying Instructor said, "I think it would be better for all concerned that you not be a pilot."

I wanted to kill everybody. "Mrs. Howell's boy being grounded, and he's one of the better fliers." I wanted to be a pilot and I thought, "Gees, what they've done to me." I begged, "Please penalize me—do anything, but don't *shoot* me, 'cause I wanna be a pilot so badly."

I thought of kamikazi: "I'll steal an aeroplane and I'll get up to 1,000 feet, and I'll dive right through the bastards. I'll dive bomb them . . . just to show them that they're wrong." I thought I had arrived, but I felt that everybody was agin me. I never got on Harvards, and it broke my heart.

Student Pilot, RCAF: Trans-Canada Air Lines came in there in those days. They had quite a radio system, and one of the big no-nos which was impressed upon you very, very often was you stay away from the pylons, which were joined by various wires for their radio system. There was a strict order: "Do not fly between the pylons."

I came in this day and there was a wide open space to land, so I went in under the wires and taxied over to the Flight dispersal. My instructor came out and started screaming, "Do you know what you just did?" And I said, "Yes." He said, "I won't take the responsibility to sign you solo again." He was upset; he told me that I'd done wrong, and I said, "But there was lots of room." He said, "You did it deliberately?" and I said, "Yes." And that was it. I turned in my gear and I dragged my ass around getting clearance from the station.

The public address system went, "Airman Wilson, report to Flights with flying gear." So I went over and there was an RCAF testing pilot there. He had taken me up for my twenty-hour check and he said, "I don't believe this; we're going up." I said, "No. If I'm finished, then I'm finished." He said, "No, we're going up. Borrow some gear; let's go."

We went up and I did everything that I was supposed to do; we did the whole ball of wax. I had done three or four landings for him and we got back, and he said, "Well, if you *want* to go, I guess that's it. You *do* work awfully hard flying a Tiger Moth with 140 horsepower. Think how hard you'd have to work flying a Spitfire with 1,000 horsepower."

They could have *given* me an aeroplane and I wouldn't have taken it. I was finished! . . . Ego . . . pride . . . just shot! And I had been home just at Christmas on leave, and here's me—I'm going home washed out. Jesus! It's the worst thing that had ever happened to me in my life. I was shattered.

Testing Officer, RCAF: It was every young fellow's dream to get up in the air. Then after they'd gone solo and had a few hours flying an aircraft by themselves and checks by their instructor—and instrument flying, and aerobatics, and what not—the instructor would come to me and say, "I've got a pupil ready to be tested." And I'd say, "Well, what do you think?" He'd say, "I'll have to leave it up to you. In my opinion he's fair, but I don't know." And I'd take him up and put him through all the paces—right- and left-hand turns, steep spirals, aerobatics, spins, his method of recovery, put him under the hood—and then make my own decision. In many cases it felt like Hell. I hated it; oh, God, I hated to tell a guy he was finished.

Student Pilot, RAF: I'd done my aerobatics and the full number of hours and we were just getting near the end of the course when one day the inevitable Warrant Officer came in and pointed: "You, you, you, you, and you—Portage la Prairie, and you're on a navigation course." It was a case of supply and demand. The Battle of Britain had been finished; there were sufficient pilots around for the single-seater fighters; they were getting into the heavy stuff and they wanted other aircrew. It wasn't a case of being a failure; it was a case of demand—a shortage of navigators. So: "You're going as a navigator."

Headquarters Staff, RCAF: It all depended on aircraft production. If they could produce more fighters they'd want more fighter pilots, and then when they started whipping out the Stirlings and the Lancs they needed seven or eight aircrew for each aircraft and that increased the demand in any particular category. That's what set the quotas.

Officer, RCAF: Quota was part of the work of Air Council. This is where the requests for training would be considered and what was needed to implement these requests. Then, having decided what they would do, Air Council would pass the instructions on, and the individual air member involved would take care of his part of the planning through his staff. Eventually people came in under that; quotas were agreed on and went through the system.

Headquarters Staff, RCAF: These washed out student pilots were sent back to the Manning Depot, and the depot said, "Well, we need observers here, and we need gunners there," and so on. But I was more or less in charge of what went on at all the EFTSs—the syllabus of training, the drill, the classroom work, the co-ordination of the whole thing—and nobody ever suggested, "Now look here, you gotta clamp down on these guys, because we need some observers and bombaimers," and what not. If the students merited continuance in pilot training, they got it.

Instructor, RCAF: One case I had was very difficult to decide. It was a lad who didn't have co-ordination and would never be suitable for high-speed aircraft. So I put him up for a progress test, and they agreed with me. It was the CFI or the Squadron Commander that washed him out, and then he remustered to observer.

Student Pilot, RCAF: I had about thirty or so hours and they called me in one day and said that they didn't figure I should stay as a pilot any longer, that I wasn't making the progress that I really should have been doing. So they said, "Away you go to Trenton—down to the Reselection Centre." By that time they had split the old observer classification into specialties so that you either became a navigator or a bombaimer or an air gunner, but when I got to Trenton these quotas were full so I was selected to become a wireless air gunner, a WAG. It was all right with me, although I didn't feel that I had any particular aptitude for doing 28 words a minute in Morse code, but the fact that I was able to stay in aircrew kept me satisfied.

Instructor, RCAF: One chap who was washed out on my course, his father was one of the top men in the RCAF, and he wrote home to Daddy—and all Hell broke loose. "There's no way that my son shouldn't . . ." So he was *put* through the course, and on his first operational flying he crashed and was killed. It could easily have been that his father didn't show very good judgment.

THE AIRMEN

I wanted to fly! I wanted to be a fighter pilot. I wanted to fly Spitfires. The propaganda then was great, and they had "Your Country Needs You" and all these glamorous pictures, and the Battle of Britain. And of course, the Spitfire pilot was the ultimate if you were gonna fly.

I DIDN'T KNOW WHAT THE PLAN WAS ALL ABOUT

Before I joined, the only thing I knew about an aircraft was that it had to be flown and the fellow that flew it was a pilot. It was something that I wanted to be, and I think the majority of young Canadians going into the Royal Canadian Air Force went in to be pilots. I didn't know anything about gunnery, so why would I want to be an air gunner— or anything else. Piloting was all I thought I wanted to do. I didn't know what the British Commonwealth Air Training Plan was all about, and I suspect that most of us went through it and graduated into whatever classification we came to be, without really knowing what it was all about. I don't recall anybody ever sitting down and specifically delineating the things that we could do in order to arrive at an end result.

I joined by accident. Five seconds before joining I didn't know anything about the Air Force. I was going to join the Army and I walked into the Drill Hall in Kamloops, BC, and there just happened to be a travelling Air Force recruiting officer there. He said, "What do you want?" . . . "Well, I was gonna be groundcrew." He said, "You'll be aircrew," and I said, "OK." So I joined the Air Force.

In October of 1940 I still couldn't drive a car, and I couldn't quite visualize flying an aeroplane when I couldn't drive a car. I was twenty years old, and when they asked me if I wanted to be a pilot, I was lukewarm; and then he said, "Would you like to be an observer?" I said, "What does the observer do?" He said, "Well, he drops the bombs." I said, "That! I like that better."

I could swear it smelled like the main street of Dallas where we slept. It was a great, big, rambling old structure, and there must have been 300 double bunks in the back where they used to keep the cattle. And the parade grounds were about four miles away.

Brandon was so full they shipped us out to tents in the exhibition grounds, and it rained darn near all the time. It was so wet we couldn't do anything, so later in Montreal we had to take some of our time off and get a bit of foot drill. None of us knew our right from our left.

At Initial Training School our day started about six o'clock in the morning, and we went through right till five o'clock at night, then we had enough homework to keep us going until midnight. There were very few people who took off for weekends, because it was really a high pressure study situation and you always had to worry that you had to pass or you weren't going to get to pilot training. You lived this fear of not making it, and that made you work just that much harder. Everybody worked, and anyone that didn't just didn't last.

We were all there as potential airmen. Some of us wanted to be pilots, some of us wanted to be navigators, and some of us wanted to be gunners. But I personally didn't know anything about the routine of what would happen to any individual to become a pilot, an observer, a gunner or whatever, and I don't think anybody else there was any better informed. We simply knew that we would go from here to somewhere else.

Student WAG, RCAF: *We had a huge radio set. It was a transmitter and receiver mounted one above the other and it just about covered the whole front of the rear cockpit; it was a very tight squeeze getting in there with our seat pack parachutes. It was a very obsolete type of set. You changed frequency by removing one coil and putting another one in, rather than by tuning it. This sort of stuff was used a lot in peacetime before the war, but it certainly was obsolete when the war broke out, and it had only a very short range.*

We carried a trailing antenna which you would let out just the same as you would in an actual bomber later on, and our main thing was to learn wireless procedures and how to tune the transmitter in the air.

Every time you had a check flight it was like writing an examination, and you never knew what to expect. That was the time they would decide if you were going to make it as a pilot, and if the testing officer thought you couldn't do it, he would wash you out.

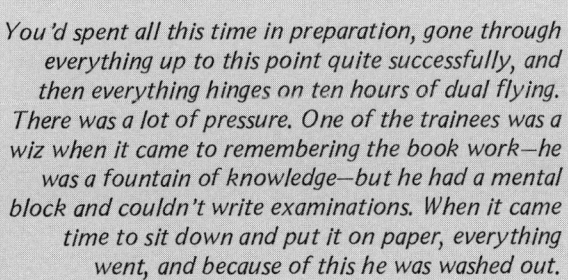

Officer, RAF: *We'd have new students coming in all the time. A class would finish and get shipped off back home to England and there'd be another class in right away on the next train to take their place.*

You'd spent all this time in preparation, gone through everything up to this point quite successfully, and then everything hinges on ten hours of dual flying. There was a lot of pressure. One of the trainees was a wiz when it came to remembering the book work—he was a fountain of knowledge—but he had a mental block and couldn't write examinations. When it came time to sit down and put it on paper, everything went, and because of this he was washed out.

I was kind of a died in the wool Brit from the word go, and I was kind of anxious to get on with it. It was a feeling for the Old Empire and one thing or another. They didn't have any call for pilots right at that time, but I was anxious to go so I put in my application as an air gunner. There was more demand for air gunners; they were disappearing at quite a rate.

Student AG, RCAF: *We were using Fairey Battles for gunnery. There'd been a change of staff pilots, and this guy, it was his first time up, so on the way out to the aeroplane he said, "Tell me when the range is right." We got up there and got organized, and I said, "Get a little closer." So he pulled in closer.... "In a bit more"—he couldn't tell. "Closer...closer," till finally I was so close I was pushing it off my gun sight, and I just filled it full, back and forth, back and forth. We landed and I'm killing myself laughing. Then they came back and said, "Wilcox, you cheated. We're not counting any of yours." Of course, everybody cheered like mad!*

Student AG, RCAF: *At Bombing and Gunnery School you had to load your own ammunition, and all the tips were marked with red or blue or green paint, and you'd haul it out to the aircraft and load it in the Brownings. When the drogue was off the ground, they signalled that it was your turn, and they had some system of knowing who was firing at a given time. Then they dropped the drogue on the runway and somebody would pick it up and they'd score your marksmanship.*

Student Pilot, RCAF: *Flying the Anson was like driving a bus. It was a very stable, sensible, sane aircraft and very easy to fly; it didn't do anything unexpected, and it was easy to land. But the first time you saw it, you were struck with its size.*

Navigator, RCAF: *The air observers got their Wings at the end of their Bombing and Gunnery School, which meant that they had completed the twelve weeks of Air Observer School and the B & G and really had the basic training for a navigator. We got our sergeant's stripes, which gave us $3.60 a day, which was really big money, and we were actually almost qualified as air gunners and bombaimers. The idea was that anybody could fudge in for any position in the aircraft, except the pilot.*

Bombaimer, RCAF: *When I marched up and that guy put my bombaimer Wings on me and I marched off the parade square, I was Sergeant Sutton—eighteen years old and a sergeant. I thought I was scared of no man and damned few women. God, I was a pretty cocky young fellow!*

3

Little boys for big jobs
Singles and twins
Sandwiches and beer
There was no communication
We learned fast, or not at all
I'd go through six or eight aeroplanes
Good pilots finished last
It was quite a feeling—an exciting day
Frustrated school teachers
You'd fly anything and everything

THE PROS

Following the completion of their primary flying course at the Elementary Flying Training Schools, the student pilots progressed to the second phase of their instruction and were posted to Service Flying Training Schools. SFTSs differed in several ways from EFTSs: these schools were administered and staffed almost entirely by RCAF personnel; civilians were no longer involved; the Commanding Officer was RCAF, as were the Chief Flying Instructor and all the flight instructors. The ground staff were mostly RCAF, and with few exceptions the ground school instructors were RCAF. In the early stages of the Plan these ground instructors had either been pre-war RCAF staff or personnel on loan from the RAF, but as the Plan expanded, the staffs on these stations came almost entirely from the ranks of the RCAF.

The types of aircraft also changed. Gone were the flivvers—the little biplanes and the Cornells. The students were now assigned North American Harvards and Yales as single-engine trainers, or twin-engine Avro Ansons, Airspeed Oxfords, and Cessna Cranes to prepare them for twin and multi-engine bombers. At the EFTSs instructors and flight commanders or testing officers had determined whether the students would be best suited for fighter or bomber operations, and the students were posted accordingly to either "single" or "twin" SFTS stations. As the war progressed the demand for bomber pilots increased, and many would-be fighter pilots found that they had to settle for the heavier equipment.

By the time they graduated from EFTS, the students were at least partly seasoned airmen, familiar with most of the ways of the RCAF . . . but they were still in for some surprises.

Student Pilot, RCAF: Bad winter weather had held up training and plugged the Service Schools, so they put us back to Manning Pool. And God forbid, they put us into a special flight and marched us down every morning to work in the kitchen—which was unheard of for chaps with a white flash. We were aircrew—we were the elite.

We took this for a few days and then Bill Batton, our sort of Father Confessor, said, "Tomorrow morning when they call for markers, I'm gonna march out, and everybody fall in on me." There were normally six flights at Manning Pool, and the thirty or so of us fell in on Bill Batton, who was marker No. 7, and everybody marched off, and *we* marched right out the front gate. Bill said, "Everybody be here tonight at ten o'clock and we're gonna march back in." We did that for two weeks and they didn't have a clue who the hell we were or *what* we were.

Officer, RCAF: I was posted to Uplands, No. 2 Service Flying Training School, as a flight commander. There were two training squadrons there and they were divided up into two flights each, and I was in charge of a flight—in charge of all the pilots and all the training of those pilots. They would come in from Elementary Schools and they'd start on their service training on Harvards.

Ground Staff, RCAF: As the aircraft came in we opened one hangar at a time, and ultimately all five hangars were operational. We would have to take the aeroplanes out in the morning and warm them up for the students. There was one heck of a lot of aeroplanes; the flight line was just wing tip to wing tip all the way down the hangar line, but it seemed to me at that time that there was a million of them.

Student Pilot, RCAF: I went to No. 5 SFTS in Brantford, and 39 of us opened that station, which was not yet complete. We arrived on November 10, 1940, in a howling wind and snowstorm, to be met by the Sergeant Major, who said, "Go find a barrack block that's got a stove in it." So we went all around and we found that there really was only one barrack block that did have a stove.

We had Mk I Ansons, and they still had gun turrets on them. They were shipped over from England, assembled in Malton, and then sent to the station. Some still had their camouflage on, and some had got worn out on Coastal Command in the UK. There really weren't many that you could call new—they all suffered from the wheezes and old age.

Junior Officer, RAF: We were finally equipped with 108 Ansons. People conjectured a lot as to why every Service Flying Training School had 108 aircraft, but the consensus finally was that somebody had said, "OK, let's say 100," and somebody else said, "Let's make it 108, and it'll look as though we'd really thought about it!" Anyway, we had 108, and I think that was normally the standard complement, whether it was a single-engine or twin station.

Station CO, RCAF: Being a Commanding Officer was a hell of a shock to start with. I'd been a businessman and all of a sudden they put me in charge of 2,000 men and 125 Harvards. We used to fly about 14,000 hours a month. I think we put eighty aeroplanes on the line every morning.

Junior Officer, RCAF: Each of these Service Schools that were built had one or two little satellites—just a place for the guys to go out and do practice landings. If you've got an aerodrome and a large number of trainees, they can't *all* practise landings on *that* aerodrome. So the normal was two: a main field and a satellite.

Ground Staff, RAF: A fitter or a rigger was in charge of the relief station, and it was usually three other men on the crew. One would be an electrician, then an aero engine man and an instrument man. You would be there for a whole week, and every morning your breakfast was brought out from the mess in Medicine Hat, which was nine or ten miles away, and of course it was cold. Lunch and supper were the same. It was a lonely week of nothing much to do. There was no proper bath out there, and being away from my wife was a terrible chore.

If the weather closed in and an aircraft couldn't land at Medicine Hat, it would land at the relief station, and the four of us—and the pilot and his student—would push it into the hangar and then make the best of the makeshift accommodation.

Student Pilot, RCAF: Our station hadn't been finished, and there was no heat in the hangar. The flight rooms were bitterly cold, so you always stood around in your flying suit. And no heat in the cans—whatever the outside temperature was, the cans were about five degrees lower, so you didn't waste too much time down there! We had no snow clearing facilities, although later we did get a plow. We'd stick evergreens at all the runway lights, the plow would go 'round them, and we'd have to go out and shovel—there were so few of us there that we did coolie work. We even did DIs on the aircraft.

Station CO, RCAF: I got the job of opening Claresholm, and we had pupils before we had runways. But we had buildings, so we accommodated them, and we had classrooms so we were able to give them their ground training. Then we got one runway, and another at the satellite field, and we used those things from daylight till dark.

We had Mk I Ansons, and you had to crank the undercarriage up by hand. We gave the students their training by force-feeding them as soon as we got the runways—working them overtime, running their training at the highest rate they could take—and we graduated them on schedule.

Student Pilot, RAF: At Calgary we got a pep talk from the CFI: "You're gonna toe the line here; there's no time for slackers. If you're no good you'll go back to be observers or something like that." Very encouraging! And then the pupils started: "Oh, Christ, there were six washed out on our course," and "Don't land with your wheels up or you'll go to the glass-house at Curry." (That's the military jail, the glass-house.) That was the start, and you thought, "Oh, Jesus, they're after me again. If you don't measure up, you're gonna be shipped off."

Student Pilot, RAF: It was a nightmare. That was the hold they had over you; they could make you do anything, anytime, because all they had to do was threaten that you would be washed out.

Student Pilot, RCAF: The stories came through: "If you think that because you got through Elementary you've got it made, you haven't. There are more people washed out at Service than there were at Elementary." And so we still had that pressure of having to work *very* hard. It was a full-time project.

Student Pilot, RCAF: They didn't swing props at Service School—you had starters and that sort of thing, so that there wasn't any hand cranking as there was with the Fleet Finches. The thing that struck me most of all on the first flight was the amount of power the Harvard's engine generated compared to the little aeroplane. And having a retractable gear was something else which was new to us—we hadn't ever had to think about selecting wheels up or wheels down—and also for the first time we had to use flaps.

Student Pilot, RCAF: *Formation flying was all just hand signals—one, two, three, go; one two, three go—by the leader. I can think of many times when I did a rate one turn in formation and by the time we'd turned ninety degrees I was two miles away from the nearest aircraft.*

Student Pilot, RCAF: We'd cruise at around 145 miles an hour, and that was great after the Fleet. You felt you were really going. You could climb up easily to ten, eleven, twelve thousand feet, and they were great to do aerobatics in, but the thing I enjoyed more than anything else was formation flying.

Student Pilot, RCAF: In those days we didn't have radio between aircraft, so we used visual signs. You really had to be careful when you gave your signals. If you wanted to go right you'd go like that, he read your message, and then on the third movement of your hand the two of you moved, literally by your hands. And if you could get his wing tucked right in tight behind, you could fly a real good formation.

Student Pilot, RCAF: We were in the vic formation, and the instructor was in the lead aeroplane with another student, and this American fellow, Derek, was on the left wing and I was on the right. The instructor gave the signal echelon left, which means that *I* drop down, move over, and come up over there, on the left. But Derek read it as echelon *right*. You've never seen such a shmozzle of aeroplanes. When the instructor looked around, *both* aeroplanes were gone, and it didn't take him too long to realize that—"Gees!"—we probably were in a heap back there. When we got on the ground he tore into Derek, but he tore into me almost as

Student Pilot, RCAF: *There must have been six hangars at Summerside, and in each hangar there must have been at least eighteen Harvards. As soon as we arrived we scrambled all over them to see what we were in for. Just by the cockpit it looked like a marvellous machine. We had to wait three or four days for our turns to fly and we watched Harvards taking off and landing and doing aerobatics, spinning, and formation flying. It certainly all looked very exciting.*

much, and I never forgave him. He gave the signal echelon left and I went echelon left, exactly like you're supposed to do. But I guess his reaction was a nervous thing.

Student Pilot, RCAF: We did night flying, which was a new experience. Later courses took night flying at Elementary, but they hadn't started that yet when I went through. Again you were looking at something for the first time—a darkened Earth with buildings and lights and a flare path—and you know that you have to put your aircraft down there.

Ground Staff, WD, RCAF: I was what they called an AW 2, Airwoman, Second Class. On certain nights I was detailed to go out and put the flares along the runway—they were the only lights they had. I stopped the truck every so many feet, lit a flare, and then set it on the ground. That was when I really felt I was replacing a man.

Ground Staff, RCAF: I was at a satellite field and I had no radio—all I had for control was a Verey pistol. If I had one aircraft coming in for a landing and another one coming in right behind him, and I figured, "He's too close," I'd give him a red flare to make another circuit.

I had seven students in the air—solo—when this awful windstorm came up and it was blowing the pots out. I'd just put the pots into the truck and light them, but by the time I'd lay them down they'd be blown out again. The students ended up all over the country—only two managed to land back at the station, and the rest of them were either at other airports or had crashed.

Student Pilot, RCAF: They were great for signal flares. If you were coming in wheels up they'd fire a red flare at you, and the same thing if you were too close behind the guy ahead—they'd fire a flare.

Student Pilot, RCAF: As the war progressed the needs for fighter pilots were pretty well filled— they were getting more pilots than there were fighter aircraft that they could stick them in—so the training switched to pilots and all the others that were required for bombers.

Student Pilot, RCAF: We all wanted to be fighter pilots, so I wanted to go on Harvards, but the word came that there was no call for fighter pilots—the people that they needed were bomber pilots: "You can forget any thoughts of going on fighters—you guys are gonna end up on heavies—like it or not!" So we all ended up on Ansons, and that was an awful big aeroplane in those days. I didn't see how the thing could fly.

Student Pilot, RCAF: The instructor said, "OK, you sit in the right seat and I'll sit in the left seat, and we'll take this thing off." And we went and did one circuit. Then he moved over to the right seat and put me in the left seat, and away we went.

Student Pilot, RCAF: You get up into the cockpit and you think, "What's this funny little thing sticking out the front?" Everything seems behind you, and to the left and right, but not much aeroplane in front of you, and you wonder, "Where did the aeroplane go?" But the fact that the aeroplane was heavier and bigger was a bit frightening at first to a young kid still just barely flying.

Student Pilot, RCAF: Almost all our instructors had trained on Harvards. I was on the second course after the Provisional Pilot Officer was discontinued, so that would have been those guys' first flying as instructors, and they were almost as green as we were. Most of them had graduated at Borden and then been given a conversion course to Ansons which had amounted to twenty-odd

minutes. In fact, they had ferried the aircraft in to open the station and then they had waited for us to arrive.

They taught us forced landings until we could land on a postage stamp, but we had very little single-engine instruction, because these Permanent Force instructors didn't know how to do it. They didn't have *that* many hours and they weren't that great as fliers themselves. My instructor informed me before the first takeoff that he had had a twenty-minute check-out, and we went through the ground drill together just in case he missed something. I tried only one single-engine landing all through Service training, and I was never given a practice engine failure on takeoff.

Student Pilot, RCAF: We were checked at twenty hours; that was your first check ride at Service. I remember the Chief Flying Instructor pulled off my starboard throttle—this was the first time I'd ever had a throttle pulled on me—and I *looked* at it, and I pushed it up again. I said, "What are you doing?" He said, "I'm checking," and he pulled it back again . . . and I pushed it forward again. He said, "I want to see how you fly on one engine." I said, "Oh, I didn't know that." I thought he was just being a smartass! I didn't know that single-engine flying was going to be part of the test, and I thought he was just trying to be funny.

Student Pilot, RCAF: Generally the flight commanders had experience. They had been in the RCAF longer, and the promotions in the Permanent Force were very rapid because there wasn't much of a body of men to draw on, so they'd been moving up. We had a flight commander we called TCA Johnson—he'd flown for Trans-Canada Air Lines before the war and he was a twin-engine man —and he knew all about this sort of thing. I hardly got off the ground when he cut an engine, and from sheer instinct I just banged open the other throttle, but nobody had ever said to do that. Then he reached over and put the power up again. But outside of single-engine work we got everything.

Instructor, RAF: There was one student who had taken a Harvard up to practise spins. As soon as he cut the engine and dropped the nose, he said there was the most awful noise came out of the engine— clattering and banging—and oil started to come out of it. When he got it back and they examined it, they found a mechanic had left a hammer inside the radial cowling, and of course it stayed put as long as he was in level flight, but when he dropped his nose for the spin the hammer tried to get out and it hit the propeller and banged around.

The Flight Sergeant came—and they're always very officious: "Airman, what's all this complaining?" He looked and he found the hammer: "Get Corporal so and so." (He had done the last inspection.) "Corporal, off to the Guard House!" Oh, it made an awful bloody mess of that engine.

Student Pilot, RCAF: On Ansons we did a lot of instrument training, a little bit of bombing training, and a fair amount of navigational training. We were constantly trying to use DR courses—getting wind estimates and course alterations—nothing particularly sophisticated, but compounded somewhat by the fact that there was no radio.

OPPOSITE:
Student Pilot, RCAF: There was no longer the fighter feeling, you weren't Billy Bishop any more. There was no longer the fun of being able to do loops and rolls and spins and all that stuff, it was just a heavy old aeroplane to fly around. On the Mark I Ansons, to get the undercarriage up you had a little short crank that you wound 137 turns, so we usually left the wheels down.

Student Pilot, RCAF: I was more frightened of my solo on the Anson than I was on the Fleet—it was like going from driving an Austin to a big truck, and there were always little lights on the panel that went on. Instructors said, "If you kill yourself in an Anson, you're better off dead!" But we didn't think it was that easy.

We used to do joint runs, where one student pilot would be the navigator and keep an air plot, and the other one would be the pilot. This was kind of fun and it kept you both tuned up; you might have known everything in theory, but it's nice to be able to do it in the air. I remember one time my partner said, "Let's bomb MacGregor Lake with a Coke bottle." And by gosh, we hit it too!

Student Pilot, RCAF: We did our cross-countrys from Dauphin to Yorkton to Regina to Winnipeg, and then back to Dauphin. They were pretty simple with the way the country's laid out in mile squares with railroads, and grain elevators that you can read names on—it's not very easy to get lost. And on our cross-countrys we used to take picnics of sandwiches and beer.

But when the final examination came and we went on our navigational exercise, they took us north of Dauphin, and there isn't a road, there isn't a railroad, no section lines, nothing. We *really* had to navigate!

Student Pilot, RCAF: It was a bit blizzardy—not too good visibility. We said, "Ah, hell, let's fly out over Georgian Bay." So we were right down on the deck. I was navigating and I said, "All right, you'd better change direction into Owen Sound." We're coming along nice and low, and all of a sudden there's a great bloody cliff! Christ! We hadn't done our map reading to know how high those hills were, and we just scraped over the top. That was almost "one aircraft missing", I'll tell you! There are some pretty high cliffs along the south shore of Georgian Bay.

Student Pilot, RCAF: I don't recall flying a buddy system on Ansons. Even on our navigation exercises we had to do our cross-country flying on our own. That was in '41; perhaps by '42 or later they did have a buddy system.

Student Pilot, RCAF: I was at Centralia just north of Clinton. I was on a fatigue instrument test and I had a brand new instructor who'd just been posted there. Clinton had 900-foot radio masts on the east side of the railroad tracks, and I had gone out under the hood on this three-hour test and I was just about at the end when the instructor said, "OK, come on out; I have control," and he was going to have fifteen minutes of flying it himself. We were fairly low and I said, "Sir, you're on the wrong side of the railroad for the Clinton towers." . . . "No, I'm not!" . . . "Yes, those towers are 900 feet and they're on the east side of the—" . . . "They're on the west side!" I said, "No they're on the east." . . . "Oh, Jesus, they are!" We just missed them; that was a close one. You can do some foolish things when you're not thinking right.

Student Pilot, RCAF: We got a big kick out of this fellow by the name of Cecil. He wasn't stupid, although he gave the impression that he was. But he was notoriously absentminded and he had no idea of what he was doing.

One of the instructors had a student up dual, and here comes a Harvard going the other way with the hood up, and nobody in the back seat. He turned and flew alongside for a while—still no sign of anyone except this guy flying under the hood. So he took the number and it turned out to be

Cecil. He was paraded for flying solo with the hood up and his defence was, "Why should I look out for everybody else, when everybody else is looking out for me?"

Student Pilot, RCAF: Everybody low flew. There used to be a little bulletin that came out detailing all the sins of people who came back with branches of trees stuck in the aircraft. But most of us were pretty safe; we had it drummed into us all the time that low flying was dangerous for even the most experienced pilot, and, "Stay away from it!" We *did* low fly with an instructor—we had a low flying cross-country trip; everybody got one. That was exciting—fifty feet pretty well all the way. Wherever there was a coulee you would fly down in it.

Student Pilot, RCAF: We did an awful lot of low flying during Service School. One time we made a cross-country formation flight with Ansons and I was supposed to be doing the flying. The other aircraft had a Sergeant Pilot with a student, and it turned out that each instructor tried to impress the other. I don't think that either of us students touched those aeroplanes at all in that exercise. Those two instructors flew so low that they had to raise the aeroplanes to get over barbed-wire fences, with the wings overlapping at times—that close. Two instructors who were both stubborn enough to have killed the bunch of us.

Teenager, civilian: One time I was coming from my uncle's place—I used to work at his farm—and it was January and I was skiing across the field. It was quite open country there, and this Anson came over. He was about 100 feet away and he was so low that when he went over the fences he had to raise up to clear them. About a mile further on there was the main power line from Lac du Bonnet to Winnipeg, and he raised up over the lines and then dropped back down. The sound was soon lost, but I must have stood there for fifteen minutes, waiting: "Maybe he'll come back."

Student Pilot, RCAF: The Anson didn't lend itself to fooling around, and the only thing that you could possibly do was low fly. And you were pretty damn careful about doing that, because if you happened to disturb a farmer he just took your number and reported you to the RCMP. They in turn passed it along to the Air Force, and they'd just kick you out. There wasn't such a shortage of people that they couldn't afford to kick a few out as an example. So nobody was that venturesome—there was too much at stake.

THERE WAS NO COMMUNICATION

They used lights for signalling—white, red, and green lights. As you were coming in on final, if you got a green light then you were cleared to land. The fellow in the control tower had sort of a sight attached to the light and he'd sight it on your aircraft, pull the trigger, and it would light up green. Even though there might be three or four aircraft on the approach, you're the only one that sees the green light—it wouldn't be lined up directly on the others. If you got a red light it meant "do not land", and you'd pull up and go round. If there was a real emergency they had flares that they would fire off—red or green or white—and each colour meant something different. Otherwise there was no form of communication between tower and aircraft until much later on in the war.

Student Pilot, RAF: I remember one of our group washing out and sitting in the barracks totally broken down by the thought of it. It was just the most shattering thing. We were so damn proud of what we were doing; we were so hepped up on the idea of becoming a pilot—not just aircrew, but a pilot—that's what we wanted to be. There is an element of shame: you've written home, and your family, your girlfriend, they're all waiting for you to go home with those Wings as a fully trained pilot.

The thought of not becoming a pilot and having to tell all those people that you've failed, that's pretty heavy stuff for a nineteen or twenty year old. Once they got washed out they tried to get them off the station as soon as possible, because they were just in despair.

Flight Commander, RCAF: Any Service School student who was close to wash out I got, to see if I could save him or determine why he wasn't able to make the grade. An RAF lad had been solo on night flying and I was giving him a check, and he froze on the controls coming in to land. He just took hold of the controls, pulled the throttles back, pulled the control column back, and there we were. I said, "You're high; put your throttles on and put your stick forward," and he did nothing. I looked, and he was just sitting there staring straight ahead. Then I realized something was up, so I slugged his arms as hard as I could and knocked them off the controls.

The Squadron Commander flew with him and came down livid—"You're trying to get me *killed*?" —so we washed him out.

He was unhappy, but he wasn't bitter. He was terrified, literally, and he knew it.

Flight Commander, RCAF: I washed out quite a few, but that seemed to be my job. If an instructor felt that a kid couldn't make it, then he would ask a flight commander to take him up, and we were supposed to try to figure it out. It really wasn't that hard—after all that training, you could tell who pretty quickly. We were really more than generous with a lot of those kids, and some of them probably killed themselves because they should have been washed out. Some of them didn't know if they were coming or going and were really far, far over their ability. But it was pretty severe before we'd wash them out.

Student Pilot, RAF: He was just George, and he confided in me that when he was sent out to do solo aerobatics all he did was fly in large circles. He had never turned an aeroplane upside down on his own; he would do it with an instructor, but never on his own. He was really terrified of flying. He must have driven his instructors nuts, because he was one of those unco-ordinated guys who swings his arms with his feet. But he got through Elementary and he got into Service, and George himself felt that he should probably have failed. But he got his Wings.

Student Pilot, RCAF: There were six of us on our course selected to be instructors. On my final tests I got an almost perfect mark in instrument flying and I was an average pilot in aerobatics and everything else. I was always good at airmanship, and I wasn't going to let some crazy nut run into me. Maybe that had a bearing on it, but there had to be an assessment. Somebody sat in the back seat and watched me fly, as I, subsequently, watched many students fly, and then I could pretty well tell if, although he was a bit reckless, he was a good pilot. I was always very cautious, and maybe that would make a good instructor.

Student Pilot, RCAF: Nobody wanted to be an instructor—that's the last thing anybody wanted. That was an eminently undesirable fate.

Student Pilot, RCAF: I had a frustrated instructor who wanted to get overseas. He was snarlly, and you couldn't do anything right for him. He was just a rumble fanatic.

Student Pilot, RCAF: They almost roped me for an instructor when I graduated; I was in the top ten of the class, and the top ten got commissioned. Through the grapevine I heard, "You're up for Commission and you're instructor material. If you accept the Commission you're going to end up in MacLeod," which was an RCAF Instructors School. And I said, "If I don't?" He said, "You'll probably end up a Sergeant going overseas. I said, "Great!" I took my Sergeant's Wings and got posted overseas.

Student Pilot, RCAF: *I went up with this instructor who told me to fly straight and level and keep that needle and ball right in the centre. I'd be working on that, and the first thing you know it'd be off to the left and I'd be off in a turn this way, and I'd over-correct and I'd be on the other side turning that way. Over-correcting was one of my things.*
Student Pilot, RCAF: *I'd try to look over the bloody nose of the Harvard, and you can't look over the nose—it's away up there! And I'd take off with the tail too high, and then she'd zig-zag all over the place.*

I'D GO THROUGH SIX OR EIGHT AEROPLANES

Instructor, RCAF: I was a keen, keen pilot—I just loved to fly. Many evenings I would skip supper and go down to the flight line and test Harvards for night flying; it was a marvelous opportunity to do what *I* wanted to do in the air instead of having a student with me.

I'd take up a mechanic. The mechanics loved this, because this was just about the only time they ever got in the air. I had a pat routine, and each test was just sort of up, checking cockpit lights and landing and nav lights, and a slow roll or two, and back in and land. And I'd jump out of one and into another, with another erk, and up we'd go. I'd go through six or eight of these aeroplanes every evening.

Ground Staff, RCAF: He'd put it through all kinds of aerobatics and that was fun for us. But I was taken up once on a real black night, and, my God, did he make me sick. He was doing loops; he was rolling; he was diving. He would bring it straight up to a stall and then let it fall back, and he let it come down like a leaf. I guess *he* was enjoying it!

Student Pilot, RCAF: The Wings test was *everything*—everything you'd learned from takeoff. Engine failure on takeoff, precautionary landings, timed turns—and it could take anywhere from an hour to an hour and twenty minutes depending on the examiner. It covered everything—under the hood, engine failure under the hood, and all the emergencies and fires—the whole bit.

Student Pilot, RCAF: I was tested by a Flight Lieutenant, and the test lasted an hour and fifteen minutes. I got the impression that it was essentially a formality; your instructor had already said that you were capable, but because of routine this man had to satisfy himself and sign a piece of paper saying, "This fellow is OK." There was not the same degree of anticipation or fear related to the Wings test as there was to finishing off the flying at Elementary—the degree of confidence had gone up, and by this time I had acquired the magnificent total of 53 hours on the Anson.

Officer, RCAF: We kept most of the good fellows we trained back to be instructors, much to their dismay. We picked the best people—the ones that would be the best to train the *next* ones—we had to do it that way. Each fellow had his instructor, and the senior staff was monitoring student progress and they sorted them out during the time of the course as to who the good pilots were and who the responsible ones were, and who weren't quite so reliable but were good enough to fly. Much to their chagrin, the good ones were told, "Your next assignment will be to Flying Instructor School." After doing their bit—a tour in flying training—they then got their turn overseas.

Flight Commander, RCAF: When I graduated out of Service School, I had a total of 74 hours and 40 minutes, but the changes in training were very rapid. I had instrument flying and night flying in Service School, and I had more instrument flying and more night flying at Instructors School in Trenton. But the first student I took up night flying at SFTS in Calgary had more night flying than I did, and the first one I took up on instruments had more instrument time than I did. Things had changed that dramatically from the Plan's start-up. I didn't want the job of flight commander and I didn't like being an instructor. I wanted to be overseas, and I was just typical of practically every other instructor.

Instructor, RCAF: For the most part, the selection of instructors was made locally through recommendations from the student's own instructor to the flight commander, to the Chief Flying Instructor, and then it would go to Air Command. There were several Air Commands established across the country, and the area command would then determine who were going to be instructors and who were not. Normally this would be done on the straight recommendation of the flight commander who had the closest all-around contact with the students.

They took the best pilots for instructors, because they felt that they would make the best instructors, but that was not always the case. Many of them were so dedicated to getting overseas that they in fact made poor instructors, simply because they didn't *want* to instruct.

Nevertheless, often without regard for the young student pilot's hopes and aspirations, decisions had to be made as to which of the graduating students would serve their time instructing. For the moment, whether they were twin pilots who had trained on Ansons or Cessna Cranes, or single-engine pilots who had gained their experience on Harvards, their formal schooling was complete. The moment they had all waited for was at hand, and with the presentation of their Wings they could feel that they had arrived. The big occasion was usually heralded by something of a bash.

Student Pilot, RCAF: After our Wings party, my buddy and I were going back to our barracks, and all the officers' shoes were lined up outside their doors for the batman to shine. There was a fire extinguisher on the wall, so we took it and filled all the shoes. The next day on parade, oh, I was hung over. I remember I fell in eating a tomato sandwich that somebody had scrounged for me. The officer was giving the drill on what to do on graduation parades and so on. His first remarks were, "Before we get into this, I just want *you* to know that *I* know who the son of a bitch was that poured the water in my shoes last night!" That doesn't help a big head on a big day like that.

We'd gone downtown and bought RAF Wings, looked at them all in the shop window: "What kind of Wings do you want?" They'd issue you a pair, but you wanted a better style, their's were kind of cheap. I remember going along Seventh Avenue in Calgary. We went in and examined them all, and bought them for our battledress and bought them for our best blues.

Senior Officer, RCAF: People who were invited to present Wings would either be a senior officer or somebody from civil life, and in some cases they would invite the father of some lad who was getting his Wings.

It was a normal Wings parade up to the point where my name was called and I stepped forward, but much to my surprise my mother appeared out of nowhere and pinned my Wings on me! To this day I don't know how come my mother and father were there, but my father was an officer in the Royal Winnipeg Rifles, and maybe that was a reason. I wasn't that old, and it was a real shock. I was really surprised.

Student's wife: I was just so excited. It was terrific to see all those fellows up there, and they all looked so smart. I felt very, very touched . . . and so proud of him. And when they came up they had to take that white flash out of their cap and hand it over when they gave them their Wings. That was a pretty proud day for me.

There was an atmosphere of pride, that you'd accomplished something. My parents were there, I think thankful that I was still alive. They were never keen about flying, but of course they'd never flown. The number of people who'd flown prior to the war and early on in the war was a pretty small proportion of the population. So it was something quite unusual and exotic for them.

The Duke of Windsor presented our Wings to us. He had been staying at his ranch not too far from Claresholm and he had a very short conversation with each of the fellows that came up.

My mother and two sisters were attending the Capitol Theatre in Winnipeg one evening, and the news came on showing a graduating class of airmen. The camera picked up their number two son, which was me, and followed me from the time I stepped back from the lineup, down, around to the Duke of Windsor, the salute, the presentation, the handshake, the turn around, and then back. And that made their 25-cent show worth a thousand dollars!

I had accomplished something that I had wanted to do. I was three-quarters of the way there. I had one hurdle yet, and that was to get posted overseas, because we still didn't know if we were going to be instructors, but on that particular day that wasn't foremost in my mind. You're just nervous; you're a kid, 21 or 22 years of age, and it was very exciting.

I'll never forget some son of a bitch of a group captain stood, and as he pinned the Wings on me he said, "How many Germans are you going to kill?" I'll never forget that. It was the ugliest thing I've ever heard. I looked at him and I thought, "You must be some asshole to say a thing like that." I was so flabbergasted I didn't say a thing. I saluted him and I thought, "Well, I'm saluting the uniform; I'm not saluting that son of a bitch!"

A fellow came and took your picture. You gave him two dollars and he said, "Where are you going?"—and you gave him your address—"I'll send it." It was a scam. I never saw a bloody picture. I don't think he had a film in the camera. "Here's a chance. Fifty suckers all leaving here in a day." I'm sure that's what he did. He was an airman, but I never knew of anybody who got a picture, and I wanted a picture to send to my mother, but I never got it. Bugger!

Officer, RNAF: I was a Lieutenant. I was the senior student, and in Hagersville I had to lead the graduating class of about fifty fellows. There were only eight of us Norwegians in the group, but it was our Commanding Officer from Little Norway that came to give the Wings, and I had to command the class. I had never had any training in giving the commands in English, and I had a special tutor for quite a few times before the parade to see that I could come out with the right commands for starting and stopping, and turning left and right. I managed!

Station CO, RCAF: I had a few Wings parades of my own. It was a great thing to have all the student pilots there and all their parents if they were able to come, to put them all on parade and pin the boys' Wings on them. And shake his hand and say, "Well done, and the best of luck." It was quite a show—it was a really nice thing.

I remembered my own Wings parade, which wasn't too long before that. The Wings were symbolic, and it really meant something to be able to pin them on a boy's chest and say, "You've done the job; you're qualified." It's an emotional experience.

You were pleased to be part of it and to go off home and show your flying badge that you had on your chest. You were pretty proud to wear that flying badge; it was a great accomplishment for most of us.

We went back into the hangar when all the guests had gone, and they read out the names of those who were being posted, and who were going to be sergeants, and who were going to be officers. And for officers they'd give you a white band to put on your ordinary airman's tunic until you got your officer's uniform.

I remember gong to the Officers' Mess for breakfast the next morning. I felt kinda out of place going in there. My instructor came over and said, "Come and sit down here." But it was funny to see all the instructors and officers, and I thought they kind of looked askance at us and thought, "Oh, God, look what they're letting into the system now."

Station CO, RCAF: In January '41 we didn't have a band or the Air Force March; we didn't have the formality or the high jinks or anything like that. We paraded on just as one group and there we were. I don't remember the ground crew being on parade even. The CO pinned the Wings on us, and we saluted and just ambled over to the Officers' Mess. But as time went on we tried to make Wings parades impressive—it was good publicity.

That parade was the highlight of the whole thing, and we were all presented with our Wings. You figured, "You've finally made it!" It was so tough there were times you didn't think you ever would, and then all of a sudden—bang!—there you are. And you're suddenly a sergeant or a pilot officer—it was great! Of course you know that automatically you're going to receive thirty days leave and you'll be able to go home and have fun.

I was awarded my Wings while I was waiting to come back from overseas near the end of the war. I didn't have a parade, but I was officially awarded

IT WAS QUITE A FEELING—AN EXCITING DAY

Prior to getting mine, I wouldn't *touch* a pair of Wings; I wouldn't put my hand on them. It was sort of superstition.... I wanted those Wings *so* badly.

Up until you get your Wings you never really feel that you have earned the right to call yourself a pilot, you're only a pilot-in-training. But when you get those Wings pinned on your chest, now you are a pilot, and it's a really proud day.

Public Relations Officer, RCAF: The thing we tried to do with Wings parades was to make them local affairs, because this was where the people in *that* town—even though they didn't have children there or their sons weren't fliers—a lot of these boys became *their* boys, so to speak. They came to belong. They would be there for two months or maybe more, and the people would take them into their homes for visits and supper and so on.

The whole station was turned out for Wings parades. The aeroplanes were all lined up in sort of a square, and we were all lined up according to Flights. Each pilot was called forward and congratulated, and the Station Commander would pin the Wings on. And there was a band, and saluting back and forth.

As each graduating class went through for their Wings you were one of the group standing on parade, so you've done it seven or eight times before and you know what it's all about. Each time it's getting closer to when you get *your* Wings, and you look at those guys like they're gods. "Imagine, getting his Wings. Gee, if I could only touch his sleeve."

You could invite guests; we got a fancy invitation and you could give it to your friends. I was one of the few that had any outside guests, most of the boys didn't know anybody, but it was nice the local people came. It was a great occasion. The band played the Air Force March, and we all marched off like we'd never marched better in our lives.

them by Group Captain Massey. It was all government routine that they were delayed, but finally they came through. He just handed them to me—I already had a set sewed on—but officially he shook my hand: "It's official now. You've legally got your Wings, as of 1940!"

*****

For most of the newly trained pilots, their Wings parade signalled the start of a period of leave followed by an overseas posting. For some it meant a posting as a staff pilot to another training school within the Plan, but for a selected few, it heralded the generally unwelcome task of teaching new student pilots how to fly. For instructor's training they were posted to Central Flying School, Trenton, or to one of the other Flying Instructors Schools which were opened as the Plan developed.

Trenton was not only the home of the Central Flying School; it also became the temporary posting for washed-out pilots and other remustered personnel. As such, it was known as KTS, the Composite Training School, with a "K" to avoid confusion.

Washed-out Pilot: Everybody dreaded Trenton—KTS. The things we heard about KTS were just terrible: "They don't give a damn about you; you'll rot there for weeks; you're a wash-out, you're finished." Anyway, I had two weeks leave at home, which with being washed out was a bit of a drag, and then took the train down to Trenton. Of course, there was more than just myself travelling to KTS; there was a whole bunch of us wash-outs.

KTS was the biggest station I'd ever seen in my life. Trenton was a pre-war Permanent Force station, and KTS was a big part of it.

Washed-out Pilot: My name started with a "W", and the pay parades at KTS were so long that my buddies all had their money spent before I ever got mine. I went on pay parade one day, found my place in the line, left the line-up, hitchhiked a ride into Belleville, picked up my boots that were in the shoemaker's, hitchhiked a ride back to the camp, and went back and got in the pay parade, and they still hadn't got to my name. That's how long it was!

Early Instructor Graduate, RCAF: We were taught a patter, but early on we were using mimeographed sheets of foolscap paper and we were tickled pink later when a small patter volume came out, because it meant that we were going to get greater uniformity. Until then we were always arguing points, within the flight and within the station, on what was the best way to instruct. The little volume was a bit scratchy in places, but on the whole it gave a lot more uniformity to the methods of teaching.

Instructor Trainee, RAF: This so-called patter tells the instructor how to explain to the student how the manoeuvre that is being carried out is actually

Instructor Trainee, RCAF: At CFS Trenton you were expected to be able to jump from one type of aeroplane to another. At this point none of us had any twin-engine training, so eventually we were all checked out on Ansons and Cessna Cranes.

accomplished. It becomes the standard, so that all instructors use the same method for teaching students, and so instructors and students can then interchange without the student becoming confused by different ideas.

Instructor Trainee, RCAF: We flew with each other for two hours at a time. One hour you'd do the patter as the instructor, and the other hour you'd be the student. You had to be able to co-ordinate what you were saying with what you were doing. If you would say, "Now I'm applying bank with a little left rudder," it was no good having the ball flying off to one side when you're saying, "I'm keeping the ball in the centre." We did a lot of flying there, and I learned to fly *accurately* both day and night, but I still thought "God, can I get a guy from scratch and teach him to fly?" By that time I'd only done a little more than 200 hours.

At Trenton we flew all kinds of aircraft. We flew Cornells; we flew Tigers; we flew Fleet Finches; we flew Stearmans; we varied every hour of the day. You could learn to fly anything.

Instructor Trainee, RCAF: We were supposed to be out practising our patter, but those Fleets were like old Sopwith Camels, so we'd get out and chase each other up and down the skyways and have a hell of a good time. We'd get on top of the clouds and do a hell of a lot of foolish things I wouldn't do now. When I think back to how we flew in those days, I wonder I'm still here.

Officer, RCAF: The quality of instruction kept improving, and the number of hours both the instructors and the students got kept increasing. We were rough on the fellows who weren't doing a good job instructing, and you got some that tried hard but couldn't teach.

Instructor Trainee, RCAF: Ernie and I went to Instructor School together. We had no intention of becoming instructors; we wanted to go overseas, but at Trenton they are up on all the tricks. They know most of the guys want to get overseas and they know damn well if you're just going around being sloppy at flying or doing it deliberately, and they don't pay any attention—they just keep you at it. But Ernie and I managed to convince them that we were actually trying, desperately, and that there was nothing we wanted more than to be instructors. That was the hardest thing to do, to appear to be as cute as mustard and to be really trying hard to do things we could do with our eyes shut anyway, and still mess up the patter *enough*, but not so much that it was obvious we didn't want to be there.

Instructor Trainee, RCAF: I was really upset when I was selected, and made up my mind that I was going to wash out of the Instructors School. This damn patter—I decided I wouldn't do it and I thought, "They'll send me to Bombing and Gunnery School," where I could at least be a staff pilot. But this silly guy put me as "above average instructor", even though we had a quarrel about the fact that I wouldn't, and couldn't, give the patter. I guess we were both young and stubborn. So I got sent to Dauphin as an instructor.

Graduate Pilot, RAF: I wanted to get my hands on a Hurricane or a Spitfire. I wasn't the slightest bit interested in being a flying instructor, but I managed to find myself at Central Flying School, Trenton. The system at CFS was that you first went through for training as an Elementary Flying School instructor and when you had satisfactorily completed that, then you went on to learn the methods for Service Flying. I didn't enjoy Trenton.

Instructor Trainee, RAF: They had a Harvard there which was a peculiar hybrid. It was a Mk I which had been fitted with a Wright Cyclone engine of something like 890 horsepower—it was a fast monster. They let me fly it one day; in fact, all I had to do was apply for a check out. I did a few circuits and bumps in it; it didn't fly like a Mk II Harvard. I honestly don't know why it was there, but they had a number of odds and ends of oddball aircraft around Trenton at that time which they seemed to be experimenting with for one reason or another.

Instructor, RCAF: The great thing about instructing was, *that's* where you learned to fly. That's where you saved your neck because you went through all sorts of situations. And just the fact that you were flying and adding to your time in the air and watching—you gained confidence. You got so you could really handle the machine. Guys that got their Wings and went overseas with 200 hours and got into a Lanc and lost an engine or got into some other trouble—they just hadn't got the air time, and that's what counted.

THE CIVILIANS

Management, Industry: *Macdonald Brothers had a contract with Federal Aircraft to assemble the Mark V Ansons. We built the wings ourselves and then the fuselages were shipped in, and the engines and so on, and they were all assembled here in Winnipeg and test flown from here.*

Official, Dept of Munitions and Supply: *Federal Aircraft was a Crown company; it did not in itself manufacture, but it placed contracts for various parts of the Anson V with a number of Canadian companies. De Havilland was one, and Boeing, which had a plant in Vancouver, was another, and Massey-Harris. These companies each had a part that they were manufacturing, and it was the responsibility of Federal Aircraft to see that they got assembled.*

THEY CAME OUT OF NOWHERE

Staff Pilot, civilian: I left the Robert Simpson Company to go to Dominion Skyways in Toronto; they opened an Air Observer School at Malton in the summer of '40 and I went with them in November. I got a private licence through the Toronto Flying Club in 1936 and by then I had run up about 350 hours. Dominion Skyways had a chief pilot, and I was checked out on the Anson, but a lot of fellows had no licence at all. The AOS didn't require a licence; they'd take anybody that had any flying experience at all. It was a war effort, like war work in a factory.

CFI, civilian: We had a great many Americans who came to us as instructors. They'd come up with certain qualifications, and if they were considered suitable they were hired on.

Ground Staff, industry: I was still going to high school, but I finished my Grade 11 and thought, "Well, that's enough." I went down to the employment office at Aircraft Repair—they were hiring. I had taken some woodworking in school, and they said, "Well, you'll have to take a test." So I had to saw this and plane that and do a few little things on the bench, but I got hired on. I was repairing Ansons, putting new plywood skins on the wings and replacing ribs. It wasn't too long before I was an inspector.

Station Staff, civilian: The elementary instructors were primarily fellows who had been taking up flying privately, and some of them had their private licence—maybe a few had a commercial licence. The Air Force had a course for them, and they had to qualify as instructors and then they were given a temporary discharge from the Air Force. They never had uniforms on.

Ground Staff, civilian: *The Elementary Schools were a gung-ho sort of operation. They wanted to keep all the people they could without having them taken into the Army, and so we joined a Special Reserve; we belonged to the Air Force, but we were on leave without pay, and when the war was over we had to be discharged just the same as anybody in the service.*

Ground Staff, civilian: *At these EFTS there were bush pilots, there were Americans, there were Canadians in the RCAF— one guy would have a uniform and another guy wouldn't. They were a mixture—as long as you could fly you were an instructor. "Prove that you can take this thing off and then come back," and you were an instructor. This guy Brown, a Yankee, came in. The boss said, "OK, take it off and see what you can do." And Lord God, Brown took off with this damn Fleet, and another Fleet was coming in with a student, and if Brown didn't land on top of him! Brown was pretty disgusted.*

AOS Staff Pilot, civilian: *Pre-war pilots are what we were. Those of us who went to the AOS at the start were all pre-war pilots; we didn't take Initial Training or anything like that, we simply had to be familiarized with the aircraft and then be checked out.*

It was a thrill to get into a twin-engine aircraft. The Anson was one of the biggest ones around in those days, and it felt heavy compared to the little single-engine Moths we'd been used to.

AOS Staff Pilot, civilian: *We were just bus drivers. We flew the aeroplanes for the navigators who had to have someone to fly them around so they could learn to do their job.*

AOS Staff Pilot, civilian: *By about 1943 the Air Force were starting to filter their own people into the Air Observers Schools. They were in the Royal Canadian Air Force and paid by the Air Force, while we were still being paid by Canadian Pacific as civilians.*

AOS Staff Pilot, civilian: *Until about '42 all the AOS pilots were civilians, but the Army was after us and wanted us to join up, so we were taken into the RCAF and then immediately given leave of absence without pay. This is the way a lot of us served in those schools throughout the war. But by 1942 the Air Force were turning out their own staff pilots and sending some of them to the AOS, so before the end of the war the majority of the AOS pilots were all RCAF people. When they joined the school they were given leave of absence from the Air Force and were then paid a salary by the civilian operator. By 1943 all the AOS pilots were in a uniform of sorts. It was blue with two silver stripes on the sleeve; it really looked quite nice. We had a sergeant pilot who got his training with the Air Force, but he wasn't operational pilot material so they sent him to our AOS. He flew as a staff pilot with our civilian uniform, but he was so proud of his Air Force Wings that he pinned them on his pyjamas.*

AOS Staff Pilot, civilian: *When we had our briefings we were given a course to go out on, usually a triangular route from Quebec City to somewhere south of Montreal, then east for a short leg, and then back to the station. We pretty well did our own navigation. We had to follow where we were going and make sure that the students didn't get us lost. And they did on many occasions, when a pilot wasn't watching or if he got into an overcast.*

Ground Staff, civilian: *I lived near Quebec City and wasn't long out of school when the war started. There wasn't much for a young girl to do in those days, but I was lucky; when the AOS opened up at Ancienne Lorette, I was hired on to help with the aeroplanes.*

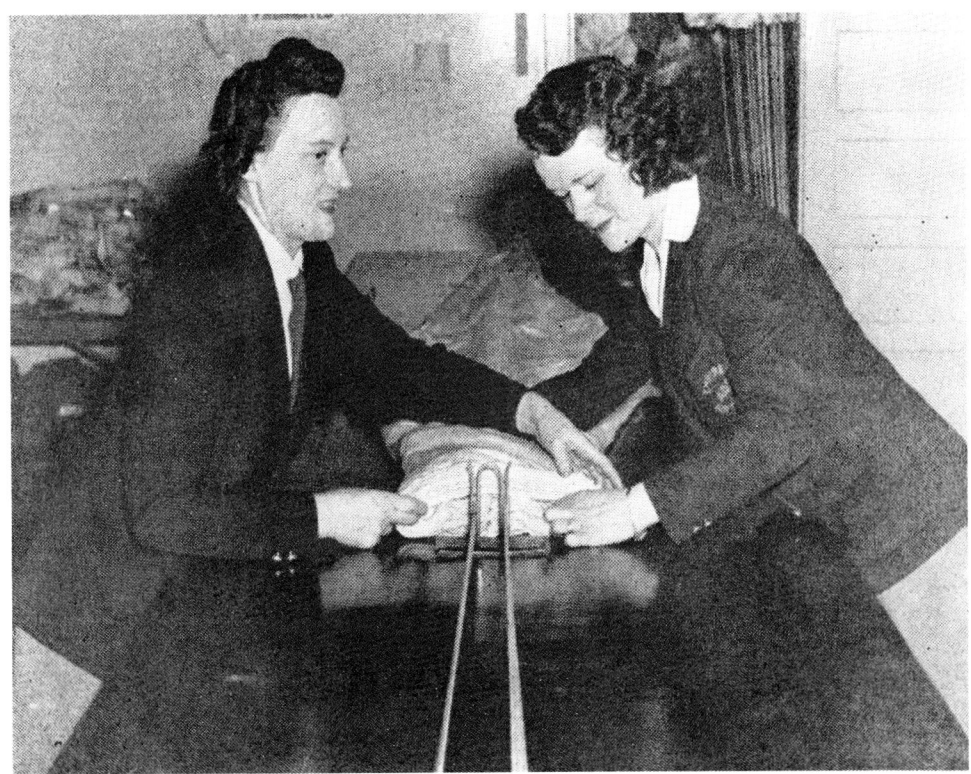

Ground Staff, civilian: *I was packing parachutes. There were five of us girls in the parachute section. Once every month those chutes would have to be hung up in the hangar and aired, then checked over and repacked.*

Ground Staff, civilian: *We had civilian girls all through. We had girls working in the mess halls and in the canteens, we had a girl who drove the crash truck, and girls on the refuelling tenders—that was cold work. The girls on the flight line were very good. They helped the pilots with their harness and when we got the Cornells they made sure that everything was clear when the pilot was ready to go. She gave the signal, and he fired up.*

Ground Staff, civilian: *We worked the radios for the navigation flights 24 hours a day. We were eight girls to a shift. WAGs in the aircraft sent us their position every half hour, except at World Series time. Then they switched over to the broadcast band and we never heard from them the whole time till they got back.*

Ground Staff, industry: *I worked on the floor as a mechanic on Harvards and Yales and Ansons. On the Yales we had a crew of our own; the aircraft went up the line in the hangar until it went out the door, and you did everything. You hung the motor, you put in the fuel system, hydraulic system, and you felt that you were part of that aircraft. If you were finished your particular section, then you'd help your fellow worker with his. You got so that when the aircraft left, you really felt that you were part of it, because you did everything but run it up.*

Ground Staff, industry: *Technically, I was an electrical apprentice, but in those days you worked anywhere the company sent you for the fantastic amount of twenty cents an hour, sixty hours a week, and no overtime. I hand-riveted ribs for the Fleet Finches, I helped assemble Fairey Battles, and I worked on the fuselages of Handley Page Hampdens that were built at Fleet at Fort Erie and sent to Malton for final assembly.*

Ground Staff, RCAF: *For a while I was a quality control officer for the Air Force and I had NCOs, sort of resident inspectors, stationed at the different plants. The Air Force Inspectors were the final quality control people before the RCAF accepted something, either a small part or a complete aircraft.*

OPPOSITE:
Shop Foreman, industry: *In the early bush flying days Macdonald Brothers on Brandon Avenue in Winnipeg were manufacturing pontoons for seaplanes. Then at the end of '39 things were sort of heating up a bit and more work was coming in. We made Anson seats and we did winter conversions for cold-weather flying to put some heat inside them. This meant building new exhaust systems to combine with cabin heat outlets. The number of workers expanded tremendously. We went from about 60 employees to over 4,000 on three shifts. There were old retired guys coming to work, and sort of middle-aged women—all kinds of them. It was a war, so there was quite a mixture, and quite a transition for a lot of us who stayed back. You'd get deeply involved in everything, because everybody was so busy.*

4

More old-time pilots
Limited separation on take-off
By the seat of your pants
We ran them until they were tired
I was young and kind of cocky
The $1.75 Pocket Ben
Hunting the enemy in the Gulf of St. Lawrence
WOGs and WAGs
Reading the headlines in Morse
Joe jobs as dull as dishwater
It was just like bowling
They're bombing west Toronto

THE SPECIALISTS

Teaching young men how to fly aeroplanes in support of the Allied cause was but one of the responsibilities of the BCATP. Many aircraft, the fighter-bombers and the heavies, were fitted with equipment that required specialized training for the operators, and often size alone dictated the need for extra hands. In the First World War the pilots handled the guns and did the navigating and located targets for the artillery, until it was decided that an extra man was required to help out. He became known as an observer, and his duties included firing guns, dropping bombs over the side, checking for enemy positions, and operating a wireless set—almost everything with the exception of actually flying the aeroplane. At the outbreak of the Second World War the category of observer was still considered as the "other hands" in service aircraft, and schools to train observers were set up across the country. They were organized by already established civilian air carriers and were operated by civilian personnel using RCAF Avro Anson aircraft. They became known as Air Observer Schools (AOS), but because of the diversity in job functions, the category of observer was soon broken into a number of different trades, each trade being given training in the other functions to allow for a degree of interchangeability within an aircraft's crew. The first AOS went into operation with the initial start-up of the Training Plan in the spring of 1940.

The Air Observer Schools

Student Navigator, RCAF: At the beginning of the war in aircrew they had pilots, observers, wireless air gunners, and straight air gunners. An observer was the navigator and he also dropped the bombs. Someone in Ottawa said, "This is too much work for one man; let's split the observer into two jobs and we'll have a navigator and a bombaimer." So they did that, but after a while someone else got the bright idea that there wasn't enough work for the navigator and the bombaimer, so: "Let's make it one trade and call it navigator/bomber." And so the navigator/bomber became the elite of the navigators.

AOS Staff Pilot, civilian. There were different flying operators for the Air Observer Schools, for example, the AOS in Edmonton was operated by Canadian Airways, in Toronto it was Dominion Skyways, and the school in Quebec City was operated by Quebec Airways. Fairly early on in the war a number of these companies amalgamated and formed Canadian Pacific Airlines, the future CP Air.

Quebec Airways checked me out on the Anson in Quebec City when they opened the school there in about May of 1941. The hangars were built, but the roads were pretty mucky, and they had to lay down wooden sidewalks. The general manager was an old bush pilot, and with him were a lot of other bush pilots who eventually became the duty pilots of a flight. They had a civilian Chief Pilot, and he had three or four as a staff of instructors. I did a few circuits and bumps with one of them in an old Anson, and away I went.

AOS Staff Pilot, civilian: The navigators were all RCAF and they had their own mess halls and quarters on the station. And if we were living on the station we had our own separate civilian messes. But in the summer the majority of the pilots rented cottages and lived off the station. There were six of us in one of them, and we bought our own groceries and commuted back and forth to the airport.

Flight Supervisor, civilian: We put out flights of thirteen and we'd have 52 aeroplanes in the air at a time. The Air Force ran the Navigation School part of it, but we civilians ran the actual operations of the aeroplanes, which belonged to the RCAF.

It was usually a three-legged exercise with two turning points, and you didn't bother with two miles separation and so on like now. We decided that at departure there shouldn't be more than three aircraft on the runway at a time: one just starting the takeoff; one lifting off; and one over the far end. The result was that they were very closely spaced, and through rather strange planning there'd be some nights when all the aircraft would be heading for the same turning points. They all cruised at the same speed, and to this day I don't know why there were never any collisions.

AOS Staff Pilot, civilian: We had what they quaintly called radios. We did what we could with what we had, but the radio equipment was certainly not one of the strong features of the operation. We had some HF—high frequency voice communication—back to our own base and we had the AN Radio Range for trying to find our way around, but failures were frequent, and if you ran into snow the static would be so bad you couldn't get either the range or voice.

AOS Staff Pilot, civilian: It was the old Marconi sets, and of course all the student radio operators were amateurs. But the equipment wasn't reliable and it particularly wasn't reliable in bad weather. If you got into rain or a snowstorm then they didn't operate, and that's when you needed them the most. So you really had to fly by the seat of your pants.

AOS Staff Pilot, civilian: The navigation trainees could take fixes, but what the hell good is a fix if it's out by twenty miles? So most of us got good enough that we were able to navigate pretty well by ourselves, and you didn't let yourself get too far off course. We had some rather crude direction-finding equipment, but there weren't enough radio stations to use for fixes. On occasion we'd have to go into the airport using the AN Range.

AOS Staff Pilot, civilian: You usually had to fly VFR—visual flight rules—to know where you were. We had our standard routes of how to get back to base: a turning point here and another there. One brewery had a miniature house on it, so that was one point that you'd pick up in the soup, and then follow 17th Avenue up to the top of the hill and turn left at the water tower, and that led you straight to base. All you had to do was see it and you sat down.

AOS Staff Pilot, civilian: In bad weather we'd fly down the St. Lawrence River until we picked up the Quebec Bridge, and then flew just past the bridge, made a left turn and flew three minutes, then made another left turn, and you're right on the end of the runway. The odd time I'd be off fifty feet or so, but you just did a bit of a turn and you were lined right up.

Student Navigator, RCAF: This old bush pilot, I remember he had a big pot belly just like a grandfather, and he had grey hair. He'd sit there and he'd be smoking his cigar, and I'd be choking sitting behind him, but he knew where he was. He flew by the seat of his pants; he could spit over the side and take a bearing. He knew exactly where he was.

AOS Staff Pilot, civilian: In the summer of 1942 the whole North was on fire. The smoke got so bad there were always ashes inside the aeroplane. We were lost in fire smoke, and I found that the five seats had only four parachutes, and I knew who the hero was going to be! So I thought, "I'll give 'er another five minutes." And I kinda said, "God, please give me a course to fly." Anyway I rolled onto a different heading, and in two or three minutes a dull glow of lights appeared, and I knew it was Leduc. So I knew exactly where to go.

Student Navigator, RCAF: One day the wind shifted and we went into the United States—right off our maps. We didn't know where the hell we were. And our pilot says, "Hang on! We'll find out." So he'd fly by these grain elevators and we'd read the names across the top. And he said, "Aw, shit. We're south of Winnipeg. We'll go into Winnipeg. I gotta get some gas, and anyway, I wanna make a phone call." So we came right into Winnipeg and we got gas, and I phoned my mother. But I didn't tell her that we got lost and had to land.

AOS Staff Pilot, civilian: We used to fly 75, 85 hours every month. We flew full; there was no waste time, and we were more than happy to be scheduled to fly three hours in the morning and three hours again in the afternoon. In that way we got our full month's hours over with fast, and we'd have the next two weeks off.

In Edmonton we were generally allowed to keep the plane we wanted, and we flew it exclusively while it was our turn on duty. If something was wrong with it at the end of a flight, we'd stay after hours and talk to the mechanics, consequently the efficiency at that station was very high.

Ground Staff, civilian: They had engine failures every now and then. After Coventry was bombed somewhere around '42, they couldn't get replacements for the engines. They were supposed to run 600 hours before overhaul, and then they raised that to 700 . . . and 800 . . . and 900 . . . and then finally we just ran them until they were tired.

AOS Staff Pilot, civilian: We all carried flares, and if you had an engine failure at night you could drop a flare and see where you were going to go in. It wasn't always motor trouble; quite often people ran out of gas. But there were very few that got killed. Those Ansons were a real safe old aeroplane in forced landings.

AOS Flight Supervisor, civilian: The best job at the Air Observer Schools was being a pilot. If you had a mean night, you'd get back and then go away and forget it all. But *we* knew which men were weak— some pilots weren't just as sharp as some of the others—and if you had 52 aeroplanes out you'd always sweat for those weak one. But there was only one fatal accident on our base. We had a few aeroplanes that landed in wheat fields, because the Ansons weren't too hot on one engine if a pilot didn't know just exactly how to handle them. But in Edmonton just one aeroplane ever killed a crew or hurt anybody. Wop May was a martinet, and everything was maintained to a very high standard.

Student Navigator, RCAF: We went to Rivers, Manitoba for four weeks. That was primarily Astro Navigation—training to be able to identify stars in the constellations and to take sextant readings to obtain ground positions. It was rather complex calculating using astrotables, and the sextants that we had were designed to take about six readings and give you an average.

Student Navigator, RCAF: It was a tough course. It's the bald prairie at Portage la Prairie. It was hot in July, and in those huts you'd have your bench where you'd do your plots and all your ground exercises and you'd be stripped naked. They had astro trainers there that you could get in and the whole of either your Southern Hemisphere or your Northern Hemisphere were there, and you'd be able to practise before actually going up in an aeroplane. It was a really good ground school.

AOS Staff Pilot, civilian: *The Anson was an overgrown Cub with two engines, a very safe old man's aeroplane. The ones in Edmonton all had the gun turrets at the back, but they'd taken out the guns. They were all glass. We called them the Flying Greenhouse.*

THEY CAME TO US FOR HELP

Licenced Private Pilot: I soloed at Barker Field near Toronto in 1937, when I was 17. We were flying Aeroncas, Taylors, and Cubs—the old 39-hp J-2s. When the war started I joined an AOS, got checked out on Ansons, and went to work as a staff pilot.

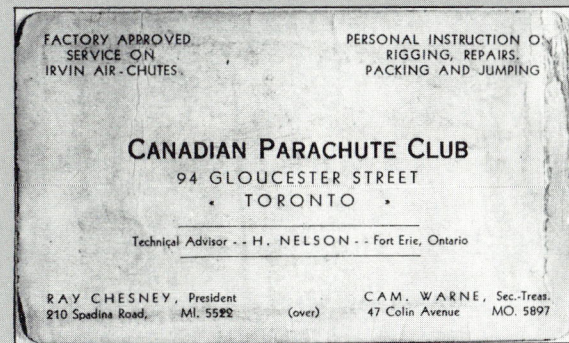

Civilian: Before the war there wasn't much doing in aviation, so I joined the Canadian Parachute Club. Then when the Air Observer School opened up in Edmonton in the spring of 1940, I went out there as a parachute instructor for Canadian Airways. They had no parachute people so they came to our club for help, and four or five of us went to the Commonwealth schools. Dominion Skyways was already operating No. 1 AOS at Malton near Toronto. At that time the AOS pilots were all civilians; in Edmonton a lot of them were bush pilots; they had experience.

Student Navigator, RCAF: We were taken up for our initial flight in an Anson with the windows open at the back. This other student wore glasses and he stuck his head out, and of course the glasses were gone. So he goes up to the old bugger flying and says, "Sir, my glasses went out the window." And the guy says, "Oh, I'll turn 'round and we'll go back and pick them up," and then he laughed like hell. Boy, were we naive!

Student Navigator, RCAF: We'd set up in crews and you'd go out flying and work on the practical application of navigation. You'd have a pre-arranged course that you were supposed to fly. Sometimes there'd be two or three navigation students in an aircraft, but there was only one desk, so one would be actually practising as a navigator and the others would be watching him and map reading. Map reading was an important part of it and it's something that takes a bit of experience before you become proficient.

AOS Staff Pilot, civilian. We simply saw that the students were maintaining track with the courses that we'd been given and mark them accordingly. If their estimates were off they were marked that way, and if they were off track that was also marked. We flew the heading, exactly what the student gave us, and if he wasn't right from one town to another, we'd mark him accordingly.

We had a bomb-sight in the Anson, and they'd calculate the winds with that. It was up in the nose and they'd fold up the right seat and slide down in there and do whatever they had to do in order to establish a wind from the drift. Then with that wind they'd calculate their course on a mercator projection on the table in the back end and come up with a course.

Staff Pilot, RCAF: They set up Air Navigation Schools to teach the more advanced types of navigation. I went to Rivers as a staff pilot. We got training in instrument flying and the radio range and extensive training in the Link Trainer. I was there for about six months on straight night flying, two flights a night, over three hours each. We had a navigator, a wireless operator, and a bombaimer; sometimes we might have two bombaimers or two navigators. We would drop practice bombs on targets at the completion of the navigational exercise. The bombaimer would take over and direct the pilot to put the aircraft over the targets.

Student Navigator, RCAF: God, it was cold! We'd sit up there at forty and fifty below shooting the stars. You'd sit there in that Anson, the thing just a-floppin' around, and dressed in these bulky clothes trying to use a little sextant. They put an astro dome on top of some of them to look through to get the shots. And you'd take readings and give them to the other fellow, and he'd plot them or look up the spherical trig. There was one fellow, he'd get awful air-sick. He'd be throwing up in some bag or something in the back and I'd cover for him, then he'd come back and take some more shots.

Student Navigator, RCAF: On some of the trips you would stay right under a hood without looking at anything except stars. If you had a position within six to twelve miles it was a pretty good shot. Shooting through those canopies, you were doing very well if you had that sort of accuracy.

The pilot was supposed to fly the courses I gave him no matter what he *felt*, so I gave him the first one and said, "I know it's not good, but fly that and I'll give you another in a moment." I took another series of shots and I knew it was out too by a few miles, but he began to shout about, "This is off," and that I didn't know what I was doing. Of course, I was young, and kinda cocky, and a little insolent too, and he was a flight lieutenant! On the third set of shots I gave him another alteration, which he refused to fly. He said he was just going to fly the beam into Brandon and he wasn't going to rely on someone who knew *nothing* about navigation.

As we were getting out of the aircraft, I said he was supposed to be flying the courses I gave him, whether he liked it or not. He said, "It's the worst navigation I ever saw!" I told him, "It's the worst bloody piloting I ever saw!" Unfortunately, I didn't report him—it was within our rights to do that. But I found out afterwards he certainly reported me for "Insults and poor navigation—giving courses that were ridiculous." But it came out in my log, "Pilot refused to fly courses."

Student Navigator, RCAF: I went to a place called Chatham in New Brunswick, and that was worse than being posted overseas because it was pretty isolated, but I went there for twenty weeks to learn the navigation part of being an observer.

Taxis were cheap. Five of us could go from Chatham air base into town for fifty cents—that was a dime each. The only thing to do there was to go the Government Store and buy Demerara rum or go to the dances in Newcastle, which was pretty close by. There was no such thing as buying a drink in a hotel, but the people in Newcastle treated the airmen great, and they drank vanilla extract cut with just about anything you could get your hands on.

There was always a dance on Saturday nights, and there was a little park in the centre of Newcastle on the shore of the Miramichi River, and a lot of our guys used to get tight and go there to sleep. They never got rolled for their wallets, but they'd wake up with their boots missing. I think everybody in Chatham and Newcastle wore Air Force boots. You'd see the odd guy coming back Sunday morning just wearing his socks.

Student Navigator, RCAF: None of us had watches. They didn't issue them in those days. So I bought this $1.75 Pocket Ben—maybe it was 75 cents—all the farmers had them. And I'd navigate by it because it had a sweep second hand. I had worked on a farm near our summer cottage and I'd got to know these Stoneberry folks very, very well, and on my final trip we were out over Lake Winnipeg and I asked for a correction. "Steer such and such for three minutes." I thought, "I'm gonna throw this watch out over the Stoneberry acres"—they had three sections there—"and when I get back from overseas I'll go and find it." And I threw it over, being quite confident that I'd hit the Stoneberry farm. I always used to write them and ask, "Have you found my watch yet?"

ANS Staff Pilot, RCAF: We got pictures. When I finally learned to handle the Anson I got my buddy out there to get a picture of me. He got out on the runway and he got me on final just touching down, and I sent it home to my father. You couldn't see who was flying it, but I was proud of it anyway. And then I got my buddy to do a touch-and-go, and I got his picture!

ANS Staff Pilot, RCAF: We had Mk V Ansons. We flew air navigation students out of Charlottetown. They were Limies—Englishmen—and they were the sloppiest bunch I've ever run into. Their aeroplanes were dirty, their hangars were dirty, and we had grey and green paint everywhere we went. It was terrible.

We were an advance guard. About two dozen of us were sent to replace their staff pilots from the Royal Air Force who had gone back overseas. We got there late at night, and they said, "If you're hungry, go to the mess." There were no utensils, no forks, no knives, no spoons—nothing. We said, "How do you guys eat?" I was red-headed at the time, and this guy said, "Gin-jaar, don't you have your mess kit?" I said, "What's a mess kit?" Every RCAF station had station utensils provided, but the RAF didn't; they carried their own. I remember that night eating fish with a penknife!

AOS Staff Pilot, civilian: There was a mix of Canadians and Americans. They couldn't get enough Canadian pilots, so they brought in a lot from the US. Near the beginning of the war most of these guys were fairly experienced. Then we got in some Australians and New Zealanders.

Student Navigator, RNAF: We were flying in the Gulf of St. Lawrence for maritime operations experience. In most cases we didn't see land; we were just right over the ocean on visual search practice. We had everything we needed to find out exactly where we were at all times. And we also had night exercises, and in those cases we had astro navigation. We picked up our own stars to shoot and we had very modern pieces of equipment, like the averaging type of bubble sextant.

When you were finished you would give the pilot the course to fly home, and if you didn't come out on the right place you haven't done very well. You wouldn't be too far out, but the pilot would know where he was anyway.

We had people there from all the Allied nations. There were Poles; there were Dutch people; there were people from France. We got into courses where the nationalities were all mixed up.

That GR (General Reconnaissance) course was interesting, because at that point there were even submarines coming pretty close to the coast, and I remember that some students did report sighting submarines.

Student Navigator, RCAF: The story was that we had an Anson lost or got badly banged up or something doing a square search exercise in the Gulf. It

JOHN GEROW   GEORGE FRANK
—Photograph by Gladys Reeves.

26-Year-Old Pilots, With Transport Tickets at 19, Like Jobs as Civilian Fliers for Training Centre

Student Navigator, RCAF: *We had to take pictures out the side window of the Anson. The camera weighed about 50 pounds, and we had to hold them just by hand. On my first trip over Toronto I was having difficulty with the pilot; I couldn't get him to settle down right on the spot. After about four tries he said, "Ah, the hell with you!" and he just turns her up on her wing tip. I'm hangin' on to this camera out the window, and I just about went out. I've got this 50-pound camera, and I'm scared to let go. Scared the hell outta me. Some of those pilots were kind of punchy.*

thought it saw a submarine and went in with its one and only 11½-pound practice bomb. And suddenly over the RT the rest of us doing our exercises all heard, "All Ansons, return to base. Do not attack!" Well, anybody who wants to attack a sub with an Anson is crazy to start with.

The Wireless Air Gunners

The civilian-operated Air Observer Schools provided the basic requirements for training navigators, and the Air Navigation Schools organized and operated by the RCAF supplied more advanced navigation training. But navigation was only one of the functions of the old observer category of airman. Another of his duties had been to act as a wireless operator, but by 1939 this had become a considerably more specialized trade than in the First World War, and it was necessary to set up a separate training program. Wireless operators were still part of the observer category, they were still aircrew, and some of their duties were those of observers, but they joined the RCAF as wireless operator air gunners (WAGs), and in most cases were not required to attend Initial Training School. Their initial training was at Wireless School, and they were posted there directly after serving their time on guard duty.

The WAGs trained at Bombing and Gunnery Schools after passing their wireless courses. Their counterparts on the ground were known as wireless operators ground (WOG) and were required to take the same wireless courses as the WAGs.

Student WOG, RCAF: There were a couple of hundred of us and we were the second course at the Montreal Wireless School. We got into the drill hall and we had to do a test on radio theory. I had done some studying on it, but I was only able to answer one question: "Do you know the difference between a Colpitts and a Harvey oscillator?"—and I answered, "No!" The other things I got right were my name, rank, and serial number. However, I'm now on the course and I'm hoping to be a WOG.

Student WOG, civilian. I had lived in Portage la Prairie all my life, and in 1942 I was sixteen years old and had just barely finished school. I don't remember what qualifications we needed, but they gave us a little test and a bunch of us girls were hired on by No. 7 AOS as ground wireless operators—WOGs. And the BCATP sent a group of eighty of us from across the country to the Radio College of Canada in Toronto. Because of the Canadian Pacific involvement with the AOS, they put us in the Royal York Hotel for the first week, and that was

Student WAG, RCAF: I was in the first class of wireless air gunners to take a course at the school on Queen Mary Road in Montreal. In fact, they were moving nuns out of the building the day we were moving in.

Montreal was a strange place at that time. The airmen especially seemed to be picked on, and if you wandered downtown alone you'd probably get attacked by some of the French-Canadian civilians. I guess because we were English, and they didn't have too much use for the war. So we always made sure we travelled in gangs.

quite a thrill for us kids who had never been away from home. At the Radio College we had an Air Force corporal to teach us procedures, and there was a civilian man that taught us the code and theory. After that first week we moved to the old Sir William Mulock home on the corner of Bloor and Jarvis and we were told, "Don't walk up Jarvis Street!"

Student WAG, RCAF: WAGs didn't have an ITS. After guard duty I was posted to No. 4 Wireless School at Guelph, which was in the old Ontario Agricultural College that had been completely turned over to the RCAF. Next door to it was the Guelph Reformatory for Young Men; there was only a wire fence separating the two camps, and those guys would be standing along the fence watching us drilling. We'd go over and bum cigarettes from them. We used to say, "The prisoners are better off than we are." We lived right in the main building of the college throughout the course, which lasted close to seven months.

Wireless Instructor, RCAF: I was teaching Morse to the WAGs. About half the guys had no intention of being WAGs, and they just sat there in the classroom and did nothing. I had to send a final test, and I was told to send it slower and slower until every one of them graduated. I got down to about two or three words a minute, but they shipped them all overseas. In 1940 when *my* class graduated, I got over 85 percent, but I never took a test in anything! I got 100 percent on the Aldis lamp, but I never saw an Aldis lamp all the time I was there.

Student WAG, RAAF: When I got to Canada we went to No. 3 Wireless School in Winnipeg. Prior to being taken over by the Air Force it was a deaf and dumb institute, and they used to say that when the Australians came over, they took the deaf out and they put the dumb in.

We were there for five months and did our radio training, and then we had our first experience in flying. They took us up in a Norseman for a familiarization flight from Stevenson Field—that's now Winnipeg International.

Staff Pilot, RCAF: In the Norseman we carried two students and sometimes their instructor went along with them. The students carried out all the radio work. We'd take them to various navigational points, but we took them where we were told to take them, and they would handle their own radio communications with their base from along the route, pinpointing their locations and calling in ETAs, estimates of their next report, and things like that.

Student WAG, RCAF: We had to learn all the radio and electrical theory and all the practical work, and how to service a set in case of breakdown. They gave us very little time off; we had to work nights and weekends. It was a real cram course.

Student WAG, RCAF: I remember an instructor teaching us radio theory. He's drawing circuit diagrams on the board with one hand and wiping them off with the other. We don't know what the hell he's talking about and we don't know what all the lines mean. It was terrible. There were a bunch of Royal Air Force fellows there who were the instructors, but there was no equipment at all.

Student WAG, RCAF: There were Australians and New Zealanders, RAF, and the usual Commonwealth troops, people from the Caribbean—a real mixture of everyone. There was a lot of competition between the Canadians and the other groups.

When we were first issued with our flying gear we were very proud of it—the helmets and flying boots—and of course we all had our pictures taken. Then we went to Menasco Moths for our actual flying training.

Aviation Historian: A Menasco engine was used in the wireless training version of the Tiger Moth, and relatively few of them were built; they were not a particularly good aeroplane. It had all this radio junk on board and it had an underpowered engine, and the results were noticeable. It really wasn't fair to the aeroplane, but they didn't care much as long as it could lift somebody off the ground and drag them along with a wireless set.

Student WAG, RAAF: We did our early operating out of the little Moths, but then as we progressed we went on to Ansons that were equipped especially for that sort of thing. There wasn't just one complete course going through the school at any one time. One would start and then move along a bit, and then another course would come in, so that there were about four courses going at the same time, each at a different stage of training, some just starting and others just finishing.

Student WAG, RCAF: We got to the point where listening to Morse and copying it became mechanical. You wouldn't listen for dots and dashes, you would just pick up the overall sound of the thing, and you could hold a conversation with someone and be able to write down Morse at the same time. Near the end of the course I was on leave in Toronto and I picked up a newspaper. Instead of seeing the headlines, what I saw was a whole series of dots and dashes. I actually read the message in Morse, subconsciously. I thought, "Boy, it's time to get off!"

The Bombing and Gunnery Schools

With the wireless portion of their training complete, the WAGs moved on to Bombing and Gunnery Schools. These, with few exceptions, were operated by the RCAF and used RCAF staff pilots and ground personnel almost exclusively.

B & G Schools used a variety of aircraft. Britain had supplied Fairey Battles for gunnery training, and Canada was building Bolingbrokes, a version of the twin-engine Bristol Blenheim bomber, which were used for both gunnery and bombing training. Westland Lysanders and Fairey Battles towed target drogues for the air gunners, and even Avro Ansons were occasionally used as bombing trainers. Almost every type of aircraft used in the Training Plan, with the exception of EFTS types, were utilized in some way in the training of bombaimers, air gunners, and wireless operator/air gunners.

Student WAG, RCAF: *I knew absolutely nothing about Morse Code, but we were learning almost one letter a day. It meant sitting in a classroom with headphones on with an instructor up front. He starts off very, very slowly, and he tells you, "By the time you graduate this course, you're going to be able to do 32 words a minute." He gives you an example, and you think, "There's no way in the world I'm ever gonna learn that!" Every day you had to practise sending and receiving and every week you had to increase your speed by one word a minute.*

Student WAG, RCAF: The classes at Wireless School were big—there were about fifty or sixty in there—but when you went to Gunnery School you didn't go in big groups, you went in bunches of twelve or fourteen, something like that. They had a number of Gunnery Schools in the West—Mossbank and Lethbridge and Dafoe—and there were others in Ontario and Quebec.

Student Nav, RCAF: I went to Fingal Bombing and Gunnery on Lake Erie just outside St. Thomas. I went to Fingal flat broke. I only had enough money for smokes till the first pay day two weeks away. I never set foot off the station for four weeks except for one day when they loaded us in a truck and took us out to a rifle range. I never went to any town at all until the night we graduated. Four weeks—Boom!

Student WAG, RCAF: They shipped us up to Dafoe in Saskatchewan, and that was a real Godforsaken place. It was away out in the prairie; it wasn't near any town or anything like that, it was just a railway siding. There was a Chinese restaurant and a little store, and I think there was a garage; that was about all. That was the main street of Dafoe, and we were there for not quite a month.

Student WAG, RCAF: I was sent to Mossbank, Saskatchewan. A very close friend of mine, Wally, was posted to Mountain View B & G. He was from Kingston which wasn't too far away, but they were splitting us up and I was going out West. My home was in Ottawa, and I thought, "Boy, Mountain View would be great, because I'd be able to whip home on weekends." So I went to see our flight commander and I said, "Wally and I are getting broken up. Would it be possible for us to stay together?" And he said, "Oh, sure. In that case you can *both* go to Dafoe!" Wally never forgave me.

Student Nav, RCAF: We got to do actual firing with live ammunition. We did air-to-air and air-to-ground and rode in the mid-upper turret of the Bolingbroke. We had a lot of fun in those things, and we used to do a lot of low flying.

Student AG, RCAF: At Mossbank they had simulators set up with a cyclorama type of thing. Enemy aircraft would simulate an attack and you'd fire back—it was pretty realistic. It was sort of a half-circle effect, and you sat in this turret and you'd get the same view as sitting in the rear turret of an aircraft, with aircraft coming at you from various angles. They showed kind of a beam of light—that was the way they accomplished it. They'd flash them at you real fast, and you'd have to swing your guns.

Student AG, RCAF: And there was air-to-ground. They would have actual targets laid out on the ground—just a painted bullseye type of thing—and we would fire down on them. The trick there was to learn deflection shooting, to allow for the speed of the aircraft and the angle of the attacking aircraft. That's really what we were learning.

Student AG, RCAF: The training aircraft were all flown by sergeants, and we were LACs at the time so we didn't associate with them. You might see

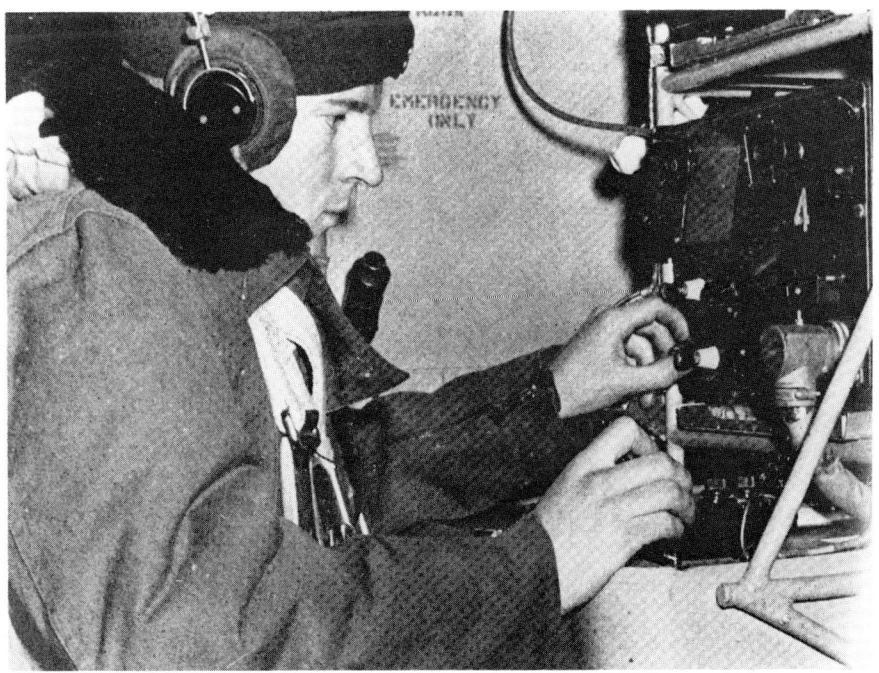

Student WAG, RCAF: *We would practise direction finding by homing in on various civilian broadcast stations, and we did navigational exercises using the old DF loop, and you had to contact Air Force stations on the ground. There was a thing that you'd turn to let out about 100 feet of wire with a lead weight on the end, and that was your transmitting antennae. Half the time you'd forget to reel it in, and it'd get caught on a snag at the end of the runway. They warned us not to interfere with the police radios, but of course we would zero in our transmitters on those things and give phony police calls.*

NINETY GALLONS OF GAS BEHIND YOUR HEAD

Staff Pilot, RCAF: My first job at B & G was to get checked out on a Lysander. It's got about 900 horsepower and you've got 90 gallons of gas right behind your head. This guy was checking me out and he says, "OK, get in the back," and he climbs in the front, and away we go. He told me everything he was doing. We were out about 40 minutes, and he came back in and landed. He climbed out and said, "Get in the front!" So I got in the front and he just stood there, so I said, "Aren't you coming?" And he said, "Not me!"

We used to take off on the taxi strips. You'd hold the brakes and get the power up, and you don't go more than 300 feet before you're airborne.

them as you got into the aircraft, but you never spoke to them. They'd have a helmet on with goggles and you wouldn't recognize them if you did see them on the ground afterwards. The only thing you got was his name so you could put it in your log book.

Student WAG, RCAF: Staff pilot was considered a joe job because it was dull as dishwater. Those guys were champing on the bit to get overseas, but they had been sent on these jobs because they hadn't behaved themselves at Service School.

There was a group of pretty wild Australians and New Zealanders who had been sent as staff pilots, and on one occasion we were coming back across Lake St. Louis just west of Montreal, not more than 25 feet off the ice, and these guys were doing push-ups with their wing tips—putting the wing tips together and pushing them up, and then breaking away.

Student AG, RCAF: Most of the ones we had were sergeant pilots who had graduated and were shipped there as sort of part of their training. I never paid much attention to them; when we weren't in the aircraft we were back in ground school. I never talked to any of them, and they never talked to me.

Student Nav, RCAF: They had these high-wing Lysander aircraft with an 800-horsepower motor. They're a great looking aircraft; I really liked them. They could slow them down with those big air brakes and flaps, and they could touch down at about forty miles an hour. They used them mostly for towing targets.

Staff Pilot, RCAF: You'd take off and climb to three or four thousand feet. You've got the drogue inside all rolled up and your drogue operator is sitting in the back. When you get to your assigned height, you meet with your training aircraft—you just pass over each other—and the guy in the back reeled out the drogue. It went out quite a ways—about two or three hundred feet.

Student AG, RCAF: They were just big sleeves. They must have been about fifteen feet long and they were towed behind the aircraft. At Mont Joli they used Fairey Battles to tow the drogue.

Crewman, Drogue Aircraft: I'd let the drogue out and bring it back in with an electric motor, and the students would play around on the outside. One time the drogue got stranded and it wouldn't pull back in, so the pilot dropped down over the river and dragged it in the water, and pulled it right off.

Staff Pilot, RCAF: The drogue had a cloth target on the end of it, and you'd fly along and these guys would make passes at it. Every now and again the gunners would get a little close to your tail, and when you had that happen: "T'hell with this! Reel in the target; we're goin' home!" And then you'd sit down and talk to the guy: "Gees, take it easy; have a heart!"

One day the drogue plane came up and I was "hose-piping"—just spraying, you're bound to hit something! I put two bullets through the drogue plane and I remember them saying when they landed, "Some son of a bitch got two holes in my rudder." And I thought, "Holy Jesus, if they find out!" A lot of the boys shot ducks; nice mallards would fly by and they'd try to shoot them down.

Student AG, RCAF: When we first went up it was in a Fairey Battle with what we called a gunner flying instructor in the back with us, and there wasn't much room. In fact, there was just one gun and it only operated 180 degrees to the rear of the aircraft because it wasn't synchronized to shoot forward. And you'd spend about half an hour or so shooting up the drogue.

Student Nav, RCAF: If three of us went up in the plane you could tell how many shots each one had put into the drogue. If mine was unpainted, there wouldn't be any colour, there'd just be a hole where it went through. I still don't know how they figured it out, but I don't think I *ever* hit the drogue.

Student WAG, RCAF: It was open to the elements; no turret—it was just an open gun position with a Vickers mounted on it; just a very simple mount, so that you could swing it by hand and fire.

We had a couple of Australians on an exercise, and the pilot lost contact with them—they just had speaking tubes—and he had to come back and land because they had passed out from the cold.

Student Nav, RCAF: One time the pilot couldn't get the undercart down in the old Fairey Battle—one wheel was down and the other was part down—so they got out the crash tenders and the ambulance and everything else. On those aeroplanes there was a wheel in the cockpit to turn the undercart down manually, but it was jammed. But the pilot happened to be a wee short guy and he was able to get his foot up and push on the handle. Then he could pull it up by hand and push it down again. So the instructor held my feet, and I hung my head through the opening in the floor telling the pilot where the wheels were. The operation seemed to take forever, but the instructor was much impressed by my apparent calmness; I just didn't know how bloody dangerous it was. We came in and made a perfect landing, and all my friends walked away disappointed as hell because they thought there was going to be just an awful crash!

Student AG, RCAF: At Mountain View we used to fly the Bolingbrokes with the little Bristol twin turret in the back. I chummed around with a fellow we called Spud. He was really built like a barrel and too fat to get into the little turret, so I used to fire *both* exercises, the one gun with red markers and the other with blue, and you only fired one gun at a time. So I would fire mine and then I would switch and fire Spud's. Somehow Spud always got a better score, and I could never understand why!

We had a staff pilot from Mexico—Mendez—and he didn't speak very good English. Well, I was wanting to get this exercise finished for Spud, and Mendez said, in his very broken English, "Ceis-a-fire." Now, if you didn't fire X number of rounds, you had to fire that exercise over, so I pretended that I didn't hear him and I kept banging away at this drogue. Mendez must have finally got mad, because he suddenly kicked it over into a turn and the next thing I know the Lysander was dropping the drogue on the spot.

When we landed there was a staff car there with the Chief Flying Instructor, wanting to know who was firing the final exercise, and of course I'd fired them both. And I'd put four holes across the tail of the RCAF Lysander—I'd almost claimed my first "shot down". I didn't meet the pilot of the Lysander, but I sure met the CFI . . . and here's Mendez screaming at me in this broken English, "When I say ceis-a-fire, I mean ceis-a-fire! Done you hunnerstan Hinglish?" Putting some slugs into one of your own aeroplanes was kinda hard to explain.

Student AG, RCAF: We had cine-camera guns to give you an idea of how you were aiming. It was a gyroscopic sight set-up, and at Mont Joli another aircraft would come in and attack you and you would try to allow for deflection. And the camera would show if you were allowing enough deflection or if you were off the target. These cameras just fit in where the guns did, and you pressed a trigger just as you would if you were firing. It would take pictures and would show you how you were doing. Our marks were recorded and we were credited with them in the ground school.

Student Nav, RCAF: I didn't think that gunnery course was very important. We used to fool around quite a bit and I didn't get very good marks. We were supposed to study, and it was mostly technical stuff, and I was with a bunch of observers and none of us studied very hard. When we were in observer courses at the other stations we had high marks and we were kind of proud of that, but when we got to Gunnery School we thought, "Oh, fiddle. This doesn't mean anything," and our marks kept slumping. We'd play poker and go into town too often, and we never studied.

Eventually this old CO got me on the carpet, and I said, "It isn't very important anyway." Well, he just blew his top! I said, "I'm not a gunner, I'm an observer." And he said, "You've got to learn guns as well as anybody else—you might have to use one."

Student WAG, RCAF: Aircraft identification was very big, more so than gunnery. If you failed aircraft identification they would wash you out of the course, and that meant all your wireless training and everything you'd had before was wasted. You couldn't even become an air gunner.

Student Nav, RCAF: I was never very good in aircraft recognition, but I was going to get through and I was on my way overseas. It wasn't going to be anything like aircraft recognition stopping me. They had a room there and on all the walls and the ceiling were the silhouettes of all the aircraft. I went into that room at night and I memorized every silhouette and its position, and when I went for my test the guy could have said, "What's the fourth silhouette from the west corner on the north wall?" And I could have told him that one and all the others without looking; I knew all their positions. At that stage I never left anything to chance.

We had training on stripping down the guns and seeing the way they worked and how they went together. I had been interested in guns as a kid— I had done a little hunting— and I was quite enthusiastic about learning how they worked.

The Fairey Battles were built like Army tanks—really solid. You could land with the wheels up and you wouldn't even scratch the damned thing.

Student AG, RCAF: The Fairey Battle was the first aircraft I'd ever been close to, and it was quite scary to start with. One gunner would be in the turret and the other would sit beside this big hole in the bottom of the fuselage, and you waited your turn.

Student AG, RCAF: There were no seats or anything in it; you just sat right on the floor. It was noisy, and rough, and oily, and everything else, but we enjoyed it. We were all young and stupid, and it didn't matter that much.

Student Navigator, RAAF: You had only two little Vickers "GO" guns, as they called them. You carried them with you, and then mounted them on a turret thing at the back, and you did your firing from there. It wasn't really a turret, it was sort of an open thing. I remember they were cold as hell.

Student AG, RCAF: They had a "G"-string that held you in so you wouldn't go into the Wild Blue. You felt that you were in World War One.

Student AG, RCAF: My mother came for graduation, and I was flying that morning, and on that last trip the undercarriage got stuck. I thought, "Gees, that's great. There's my mother sitting in the hotel waiting for the parade. If only she knew her son was up here flyin' around and we're trying to shake down the undercarriage. She'd have a fit!"

Staff Pilot, civilian: They had bombing exercises. You had to fly out to the bombing range, and the bombaimer would lie in the nose of the Anson looking through the hatch, lining up the target in his sights. In 1941 there was no radio communication like intercoms, so they'd lift their right leg up if they wanted you to go to the right, or the left leg if they wanted you to go left. When they were on target and wanted you to hold it steady, they'd lift both legs until they let the bombs go.

Student Bombaimer, RCAF: I was scared of heights, and I'm lying down in the nose of an Anson with nothing between me and the ground except that little hunk of plexiglass. But bombing was fun. It was like bowling, lying down there giving instructions to the pilot. The target is coming up towards you and you are supposed to see that it gets into the drift wires on this sight thing. As soon as the target is aimed you press the button. "Bombs gone!" It's up to you to give the pilot directions so the target comes down the drift wires, and you're hoping you can get him in position before the target disappears. Anyway, you drop six bombs and you go down and get your score. It's just like bowling!

Student Bombaimer, RCAF: They had triangular wooden targets set up. The main thing in your bomb aiming was the grouping of your bombs, that you weren't overshooting or undershooting or going from one side to the other, so if there was an error in wind speed your bombing wouldn't be affected. You bombed from four different quadrants, and if you got a bad score you blamed it on the pilot, because a lot depended on the person flying the aircraft. With the type of bombsight we were using, you had to be pretty well straight and level. Your airspeed had to be dead on, and your heading dead on, and your corrections and so on had to be good. So a lot depended on the person flying the aircraft. That's when you started to know what teamwork was really all about.

Student Nav, RCAF: I went to Crumlin, which is just outside London, Ontario. Crumlin was run by Leavens Brothers.

When you flew these exercises you rotated. One guy went in the nose and dropped so many bombs, and then the next guy went in and dropped his. When you went on to nav exercises, the one navigator took you out and another one brought you back. There was the pilot and a staff wireless operator to get radio type fixes. Sometimes it was two bombaimers and two navigators, and sometimes if it was a straight bombing exercise it was four bombaimers.

Staff Pilot, RCAF: We'd climb up to five or six thousand feet and make clover leaf patterns over the target area, and they'd drop a bomb from each direction. Then we'd switch students and make another trip. It was a pretty dull, monotonous thing, because the Anson sort of did everything at 100 miles an hour.

In Lethbridge the wind quite often blew like hell, so there were occasions when running into wind on the target would take you half an hour, and going downwind the student would say, "Left, left . . . dummy run," because the target was gone past even before you could get near it. One student was being very, very fussy—too fussy—and he kept calling dummy runs. But this particular aeroplane had an extra button on the panel, and I discovered if I pushed that button the bomb would fall. This guy was the best bombaimer in his course and he was working on getting the record for night bombing, but every time I pushed the button he got a terrible error, and *he* got blamed for it! He wasn't too happy when he found out.

Student Bombaimer, RCAF: We had a bombing range on Lake St. Louis, and they had set up a great stack of evergreens on the ice as a target, just off shore from the Caughnawaga Indian Reserve. We had circled around and we were going in on a run, dive-bombing with eleven-pound practice bombs. We went into the dive and were just about to push the tit when a bunch of kids ran out from under these Christmas trees—they had gone out to play on the ice. It was a natural place for kids to play, and we were just about to let fly.

Student Bombaimer, RCAF: We had one crash while I was at B & G. We got into a storm and lost power or something, and the pilot decided to go down into a field. Everything was fine, except instead of landing in the field he aimed for, we landed in the next field, and we went between two trees and sorta took the wings back. Nobody got killed or anything, but the pilot got smacked in the face. I was back in the rest position strapped in and I just got badly shook up, but the others all had cuts and bruises of sorts.

I remember crawling out of that thing and

Staff Pilot, RCAF: *I volunteered to fly on night bombing, but that wasn't very exciting. The only way B & G School was exciting was if all your batteries failed and your radios failed, and you had no lights. Then you could make a low pass at the tower and warn them you were in the circuit, and they would fire off green flares and red flares, and there'd be all kinds of excitement. So we used to turn on every switch we could find in those aeroplanes and run the batteries flat; then we'd come in and beat up the tower.*

remembering my drill. I put the safety pin back in the 11½-pound practice bomb and duly noted it in my log book. I thought I was something! There was only the one bomb left, and I had made that bomb safe, only to find out that I'm the guy that's stuck with the wreck, because all the rest of them went off to the hospital to get checked. As he was leaving, the pilot said, "Don't let anybody on this aeroplane until the wreckers get out here." He was a civilian pilot, but I had to say, "Yes, Sir"—but I didn't know who the hell he was—and because I was the only one that wasn't hurt I got stuck to stay with the bloody old wrecked Anson.

There I was, no weapons or anything, and I'm supposed to guard this thing and keep everybody away. But of course the people started to come out to look at it. I kept saying, "Come on, get away. Get away!" And it kept getting darker and darker, and I realized I was going to have to spend the night there. So I said, "You know, there's still a live bomb in that aeroplane!" Well, you should have seen everybody beat it; they weren't coming to look at this old beat up Anson it if had a live bomb on it!

Student Bombaimer, RCAF: I was at Malton, and the bombing ranges were just on the western outskirts of Toronto. On my first trip we had those little five-pound bombs—I think they let off smoke when they hit. We got to about 5,000 feet and I opened the doors, and I was trying to be very accurate so I got the target right in the sights, and pushed the button. Then I looked and I couldn't see it drop. I couldn't understand and I thought, "That's strange. It must have dropped so fast I didn't see it." Anyhow, I kept getting the pilot over the target, and I dropped five more, and I never did see one go.

We got back to the station, and before I got into the office one of the airframe kids came along and said, "For crissake keep your mouth shut. Don't say anything." All hell was breaking loose. Somebody had been dropping bombs all over western Toronto! When I landed, the bomb doors were open, and they'd been open when we'd taken off. I'd closed them over the bombing range, dropped the bombs inside the doors, then opened the doors again when we were on our way home and just over the edge of the city. So I'd scattered them all over west Toronto.

Two days later I was invited for dinner, and this Scots lady said, "Did you see in the paper where they're dropping bombs around us out here?" I said, "No! What happened?" . . . Oh, God!

Flight Supervisor, civilian: *The first winter was most unpleasant, because those Ansons were built in England and they had no heat in them. Some of the Australian navigators couldn't understand why their hands became stiff, and after several fellows had frozen their knees we decided we wouldn't fly if it was colder than 10 below 0, Fahrenheit. Then during the next summer they installed flexible heating pipes, and you could pull the leg of your flying suit out of your boot and stick the pipe up inside.*

Engineering Officer, RAF: *They got some thing around the exhaust which blew hot ai into the cabin through tubes. There wasn' enough heat to go around, so people woul just stuff the tube into their pants pocke and blow hot air into their vitals . . . an keep them warm somewhere.*

IT WAS BLOODY COLD

Ground Staff, RAF: At times the wind would blow the snow right up against the hangars. The doors used to slide open, and there'd be a pile of snow in front of them. The only way to remove it would be by hand. And it was bloody cold—sometimes 35 or 40 below, which we had never witnessed in the UK.

Ground Staff, civilian: In the wintertime we had to drain the oil out of the aircraft every night. Most of them were in the hangars, but it was still pretty cold in there when it got down to 20 below outside. So we'd put the oil in big barrels and keep it in one of the warm workrooms. Then in the morning before we were able to start up the aircraft we'd have to get that oil back in the tanks. Sometimes it took two or three hours before we had them all going, and there were times we only got a few of them into operation.

The snow plowing equipment was not the best, and they'd hard pack it with rollers. We did a lot of landings and take-offs on hard packed snow. No matter how much snow we'd get, the runways were ready in short order.

We had no snow clearing facilities, although later we did get a plow. But in the beginning we'd have to go out and shovel.

Ground Staff, RAF: We had a student that made a bet that he could run around the hangar naked, with just his jockey shorts. It was a stupid bet, because it was about 30 below with just a terrible wind, and he collapsed from exposure before he reached the door. He later died.

Engineering Officer, RAF: One chap got lost and landed at a farm with one of the Ansons to find out where he was. It was really cold, and he got down OK because the field was well frozen and there wasn't very much snow, but he wanted to phone the station and he made the mistake of shutting down the engines when he left the aircraft. Of course, when he got back they had cooled off so much they wouldn't start. We had to drive out there about thirty miles and warm the thing up with proper heaters.

When I heard I was coming to Canada I tried to find out, "Will it be cold?" I was told, "Probably it will get down to 15 or 20 below." I thought they meant 20 below freezing, but, by God, they meant 20 below 0 . . . Fahrenheit!

In the winter we had a canvas cover to put over the engine, and they had what they called a plumber's blow pot that they put inside this canvas thing, and you had to have somebody sitting in there with a fire extinguisher. If it caught fire, orders were, "Let off the extinguisher and get the hell out as fast as possible."

The Tiger Moths that we took our elementary training on were equipped with skis. It was just a grass field that was covered with snow and ice, and we did all our training on skis. They maintained the field by rolling it. A tractor would pull two or three rollers which had sort of a corrugated exterior, and this made little ridges on the snow. But it was flat and even.

Ground Staff, civilian: *When the frost came out of the ground it got pretty juicy, and the whole operation would move to Emma Lake, about 25 miles from Prince Albert. The aircraft would be flown up on wheels in the early morning when the ground was frozen. We had a strip cleared so they could land at the lake, but then we changed them back to skis. We used to live in summer cottages; it was mighty cold, but we managed all right. It was a bit of a holiday, really; it was different.*

Student Pilot, RNAF: *At Muskoka we had most of the aeroplanes on skis. A couple of them landed on the ice on lakes and went through, and we had to go and pick them up. Sometimes when the airport was closed because of problems after a big snowstorm, some of the lakes were in very good condition, and it was a matter of getting off at Muskoka and then doing circuits at one of the lakes. This happened quite a bit in the winter.*

Student Pilot, RNAF: *The landscape in the Muskoka area is very similar to Norway—lakes, small farms, open areas—so it was almost like being at home*

Engineering Officer, RAF: *Sometimes we'd have to take the damned Ansons apart and bring them back. One landed near a little town called Plumas up north of Portage, but we were able to make repairs on the spot. The crew went in a snowmobile, sort of a cabin thing with a propeller pushing at the back. There was a farmer there who had a team of horses; he had a railway rail about thirty feet long and he dragged this up and down the field and flattened out a runway so the old Anson could get off.*

Trainee, RAAF: *I had never seen snow until I came to Canada. I was visiting some Canadian friends one day and it had been snowing. I didn't know what to do with myself, so I asked the lady, "Would you like me to shovel the snow for you?" She said yes, so I got a shovel and shovelled the walkways, and then I started on the lawn. I shovelled all the snow off the lawn so it was clean. Afterwards she said, "We didn't like to tell you because you seemed to be enjoying it so much." They had the cleanest lawn on the street!*

Student Pilot, RNAF: *One of the jobs they gave us if we did something wrong was to shovel snow. There wasn't much snow equipment around in those days, but at Muskoka we were sort of in the bush, so horses were pretty easy to find.*

Officer, RNAF: *At Muskoka there was so much snow in some of the storms, that we had to get about 200 fellows out just tramping the snow down for the whole length of the runway—nothing but their own boots and their own weight. And then they rolled it afterwards.*

Engineering Officer, RAF: *We used to have Ansons force-landing all over the place. One landed near a little town that was completely closed off in the wintertime except for the railway, which had one train a week. We commandeered one of these trolley things with a little engine on it, and the flight sergeant and I rode up on that with two propellers, one across each of our knees. There were about forty miles of straight track with about 40-below weather, and we had to sit on this thing at about thirty miles an hour*

5

In the wake of disaster
A bowlful of buttons
Any old bolt wouldn't do
Sunshine and lobsters
You didn't feel cold . . . till after
Holey Ansons
The band played "There'll Always Be an England"
The fad from Down Under
New experiences in the snow
The first pangs of homesickness
A Norwegian summer home
Learning English with Scarlet

THE VISITORS

With the disaster of Dunkirk in the early summer of 1940, all plans for using France as a forward base for air training operation had been scrapped. The RAF had somehow managed to continue limited training in England during the summer, but with the increase in German air activity over Britain, air training became ever more difficult. The BCATP was barely getting under way when the British requested that entire RAF air training schools be moved across the Atlantic and incorporated in the operations of the Plan.

Officer, RCAF: They put the proposition to Canada and said, "Lookit, we have a problem here trying to train people to keep up our numbers in the RAF." So they transferred some of their own training schools to Canada. In those days it was a question of, "How can we get this done?" So Canada said, "Yes," and we put a big effort into it, and we then had both RCAF and RAF operated EFTSs and SFTSs.

Senior Officer, RCAF: The British started bringing their schools over. The move was precipitated because of the trend the war was taking. We had aerodromes being built to look after the RCAF requirements and the Commonwealth Air Training Plan as a whole, but not for moving in entire British schools. They really flooded the resources of the Royal Canadian Air Force, without any question or doubt.

Official, Dept. of Finance: Moving the British training schools out here meant a bigger demand for schools to be built. They brought their own training personnel and they brought the people that were going to be trained, and there were then expenditures in Canada over and above what we had expected. So the program in total—including the British Commonwealth Plan and the RCAF's own training work, which was largely folded into the Commonwealth Plan, and then the unexpected British schools—all had to be run as a more or less integrated total.

Ground Staff, RAF: We thought we were going to a war zone, but then when we got to the ship we saw Canadians on board, some of them with their arms in slings and some with crutches. So I asked one kid, "Where are you going?" He said, "I'm going home. I was in London, and they're sending me back." This is when we knew that the ship was bound for Canada, and there weren't too many of us knew that there was this British Commonwealth Air Training Plan there.

Junior Officer, RAF: We really travelled in style. The officers had the first class dining lounge and we had all the booze we wanted, and waiters and the whole bit. We used to have sing-songs in the bar every night and we had a very good time all the way over.

Student Pilot, RAF: I don't think I ever had ten worse days in my life. Four thousand of us in that boat designed for 250 passengers—everything stripped out of it. When you went on they handed you a ticket that said, "C Deck—hammock". You go down there, and there are people on the tables, people on the floors, and people in hammocks. And the hammocks were hung in a *solid sea* and touching each other as far as you could see. Oh, God! And you weren't allowed to take your clothes off; you weren't allowed to take your boots off; and there was hot water for only ten minutes a day.

Student Pilot, RAF: We got to Halifax and docked in daylight. It was a new world, completely. Here were big trains and long coaches, and bells ringing on the train—things that we'd heard in movies, but never seen. So we marched off with our kit bags and got on to this great big troop train.

Student Pilot, RAF: A whole bunch of us stood around the front end because it was the kind we had seen in the cowboy and Indian movies—a big steam burner and a cowcatcher. I remember the cars as being very uncomfortable because they all had wooden seats.

Junior Officer, RAF: We were in sleeping cars with seats that made up into double berths—very comfortable standard Canadian Pacific Railway accommodation in those days. I don't know how all the troops got along.

Student Pilot, RAF: They were long cars that had a stove at one end and a sort of cooking arrangement at the other. Hard seats, no cushions on the seats, and sleeping berths that pulled down. They had them for the big harvest trains in the Thirties when thousands of people went to the West, and they moved them in these colonist cars. The old colonists cooked all their meals on these stoves.

There were three of us allocated to a sleeping space. Two slept down on these hard seats pushed together, with just one blanket, and then one was upstairs on the thing that swung down like an upper berth. Slats—hard slats—no mattresses.

Student Pilot, RAF: At the little towns the railroad often ran through the centre of the community. We had a twenty-minute break at Truro, where we all got off just for a look around. There must have been seven or eight hundred of us on that train, and I wandered across the track, and here's a little shop with bananas in the window. I hadn't seen a banana in Britain for years. I had no money—they had taken away our British money and at this stage we hadn't got any Canadian money—but I went into the shop and I said to the girl, "I'd like one of those bananas, but I don't have any money." She said to me, "Oh, give me a button off your tunic." So she gave me a scissors, and I cut an RAF button off and gave it to her, got this banana, and came out eating it. These scores of boys ambling up and down the street saw the banana and descended on that little shop like a horde of locusts and just cleaned the place out. I'm sure that girl ended up with a bowlful of RAF buttons.

Student Pilot, RAF: I remember stopping, and a bunch of women and girls came aboard and gave us buns, coffee, cigarettes, candy. Apparently they met every troop train coming and going and gave this type of hospitality. There was a great interest on our part: nineteen-year-old Englishmen, and you've never seen a Canadian girl before. I remember the cars sort of being *invaded* by young women.

Airman, RCAF: I had a little prejudice against English people because we were very sensitive about being called colonials—that was fighting words, and that's why they had trouble with some of those English boys when they came over. They'd call us colonials, both Canadians and Australians training with these RAF types. There'd be terrible fights, because "colonial" was a dirty word.

Senior Officer, RCAF: When it was decided that the RAF schools should come over, the first to arrive was sent to the Service Flying Training School that was completed the most, and that was Carberry in Manitoba. They arrived there in the fall with unfinished buildings and unfinished runways, just in the mud of the West. After that they flowed in very fast, and it taxed the capacity of the RCAF to have them.

Junior Engineering Officer, RAF: We were very astounded. They'd built the place after they'd taken the harvest off in the fall, and they'd built just about the whole damn station from September to December. It wasn't really very well finished,

We started to put our Ansons together and found we were missing a handful of bolts or nuts or special screws for each one of them. When they took them apart in Britain they had thought that any old bolt could fill in, but being British threads, you couldn't find them in Canada. We finished about three of them and got them flying, but the rest we just gave up on; I don't know if they ever flew again. (They had all been flown in England.) It shortly came up that the arrangement obviously wasn't going to work, so they started a system of service depots, and as other aircraft came in they were all put together in a big assembly plant.

and they had to put us up in the station headquarters to begin with until the Officers' Mess was completed. But the hangars were all completed and heated, and the hangar doors were working. We were very impressed with the big steam boilers. The whole thing was a very efficient operation that they'd managed to get going in only three months.

There were two of us Junior Engineering Officers. We were only 24 years old and between us we had about 600 men that were directly responsible to the two of us, which was quite a responsibility. Fortunately, we had a Chief Technical Officer and he had a Technical Adjutant and somebody else in his headquarters. Then there was the maintenance wing, which was my own little world.

All we had were six Ansons in packing cases. They had come with us and they showed up when we showed up. The rest were coming along, and the idea was that we were to put these first ones together and fly them.

The Canadian construction teams quickly caught up with the requirements for the RAF schools, and the British stations were integrated into the operation of the Plan. The doors were opened for a flood of British trainees, and this would continue for two years.

Student Pilot, RAF: We were told, "To Canada," so I knew I was coming to Canada. On the ship some of the ports wouldn't shut and the water came in. There was a burst oil pipe, and we broke down about two days out and just wallowed for a couple of days. Every time the flat bottom came down on a wave we thought a torpedo had hit, and it was agony downstairs. We had hammocks, and as she went forward you could hear all this water and stuff swilling down, and you'd pull yourself up and the whole mess would go underneath, and you'd settle down again, and then she'd roll back, and the stuff came back again. When we landed from the *Louis Pasteur* I swore that I would never set foot in another ship. Every person on board, from the captain to the cabin boy, was ill. We were just lying around on the deck, green and white, and it was horrible. That was the worst part of the war.

Student Pilot, RAF: They *never* told you where you were going. We thought it might be Canada or it might be South Africa, but wherever, one good thing about the troopship was the food. They had butter and milk and white bread—that we'd never seen in Britain for years—and eggs! Eggs were a thing of the past by this time in '41. You had a little ticket, and at every food call you went to the galley with pails and things to get food for twenty people, and they'd load this up. When you got back to the table you found that some of them had faces the colour of greenback dollars.

Student Pilot, RAF: We had heard all the stories about the North Atlantic, and in '42 it was still a hot place. We were all alone and that surprised us, but according to the Merchant Navy people, the ship was fast enough and at 25 knots it would have to be a very tricky U-boat to sink us.

On board were a bunch of German prisoners who were being shipped from Britain to Canada—*Afrika Korps* people and some U-boat crews—and they weren't allowed up until we were three days out. They had been taught that the Atlantic was a German ocean, that the U-boats ruled this ocean, so we were all agog when these guys were brought up. The first thing they all did was run to the rail and look around, and they joined in sort of an animated conversation about the fact that there were no escorts. It must have been an odd feeling for them, believing that no ship could cross without being sunk.

You look across and you see a young German *Afrika Korps* guy not much older than yourself looking back at you, and you realize that if circumstances were just a little bit different you would be trying to kill each other. They were playing mouth organs and singing rather sad songs, and it was a strange feeling looking at your opposite numbers that closely.

Student Pilot, RAF: When we arrived at Montreal we were being towed in, and from the tugs they were throwing up apples and oranges and everybody was reaching over and trying to catch them. We hadn't seen those for a long time, and there were howls of derision when they fell back into the water, and shouts of triumph when one was caught and the chap would disappear munching on an apple or an orange.

Student Pilot, RAF: The train was an oldie, and we were really, really crowded. You slept on the seats sitting up. I don't think it was ever intended to be a sleeping car, because I can't remember there being any sleeping accommodation. It seemed we went for days through the forests of Ontario, and then we hit the prairies.

Student Pilot, RAF: The seats could face each other; the backs swivelled back and forth. They had pot-bellied stoves at the end of each car, and we just wrapped up in our greatcoats and kit bags. It was very, very sooty and uncomfortable.

Ground Staff, RAF: We arrived in Regina about two o'clock in the morning, and there were thousands of people there to meet the train. They had hot dogs and chocolate bars and popcorn and bottles of soft drinks and cups of coffee. And *urns* of tea, because they knew we were English. Until then I didn't realize how big Canada was and that you could fit England into Alberta five and a half times.

Student Pilot, RAF: At that time in '42 the Royal Air Force routine in Canada was to send everyone to Moncton in New Brunswick for initial posting. There was no transport at the station, so the bunch of us had to march up to Moncton Postings Disposal Centre or whatever they called it. The place was about nine-tenths empty. It was just a new station then, in the winter of '41/'42.

Student Pilot, RAF:
Whenever there was a whistle stop or the engine had to put on water, in places where there was hardly a platform, there would be people with baskets of fruit and sandwiches and drinks. Half a dozen people here, twenty there. And when we got to Winnipeg there was quite a contingent set up with little gift boxes of sandwiches and chocolates.

Student Pilot, RAF: No. 31 Personnel Depot at Moncton was a great, big, hutted camp and it must have held 10,000 people. The numbers of all the RAF stations in Canada—whether it be a supply depot or a personnel depot, a flying school or a gunnery school—they all started with a "3". So instead of No. 1, Moncton was No. 31.

Student Pilot, RAF: The camp hadn't been finished, and we worked on roads and dug ditches and things like that, but the food was great. Big, beautiful, clean cookhouse, compared to the RAF cookhouses at home where they had *awful* stuff. Here you sat down with tablecloths, and selections of jam and cheese, and milk and butter and eggs, and white bread—white, white bread. In Britain we carried our irons all the time—you used your own knife, fork, and spoon—but here you picked them up and cleaned them, and put them down again.

Student Pilot, RCAF: We had to wait nine weeks in Moncton, but it was no sweat. It was in the summer, and we went down to Shediac, which was a lovely beach about twelve miles out, and spent all our time sunbathing. I never was so brown in all my life—continuous sun and lovely warm water and sand. The lobster boats used to come in there, and the fellows would throw one up to you for 25 cents! Beautiful lobsters—oh, God!

Student Pilot, RAF: There was a bus that went from Moncton down to the main gates of the camp, and it was always filled with airmen and officers. You would be going home in the dark, and suddenly the bus would stop at what seemed like nowhere, and the driver would call out, "Hole in the fence!" All these guys that didn't have passes would get off, and the officers would be looking out the window as if they didn't notice. It was a regular stop and there actually was a hole in the fence there, and the guys would get off and go through and the bus would carry on.

Student Pilot, RAF: The Moncton locals were a little unfriendly. The French-speaking people weren't in favour of Canada going into the war, and there was quite a lot of animosity shown to the service, particularly to the RAF personnel. There wasn't much unpleasantness, more of an aloofness, rejection—it was noticeable. If you went

Moncton was the only place where I was ever kicked out of a camp and told to stay out until night time. They didn't have enough staff to look after us, there were no training programs there and there weren't enough recreational facilities. The Corporals were disinterested and they'd march you up and down and then they'd march you out of the camp and dismiss you. They told us the CO's inspection would be done and they didn't want the place messed up and they didn't want us lying around in the dormitories, and so, "Bugger off! And don't come back until after supper!"

We'd heard that there was a liquor store close to one of the stations that we stopped at, so we went down there. In those days there seemed to be nothing to stop us, so we took our various supplies on the train. We found that going across Canada there were only two places with liquor stores close enough; one was Riviere du Loup and the other one was Brandon. So we stopped at both places and got something to drink. In those days Canada was pretty dry.

into the shops there was a definite coolness; you would be served, but that was it. Often you spoke to somebody and they just sort of walked away; there was no response.

Student Pilot, RAF: From Moncton you just got these postings—"Swift Current, Saskatchewan"—and you'd wonder, "Where the hell's *that?*" We found Swift Current was a small town. Everything was clapboards and hitching posts, and horse-shit down the street. "Good Lord, what have we got into? How far west have we come?" The mystery wasn't too long in solving: we happened to arrive right in the middle of what they called their "Frontier Days", and everybody lines their horses up and down the main street, grows a beard, and carries a six-gun.

Student Pilot, RAF: We'd just arrived in Neepawa, and it was very cold. We'd missed the bus and the camp was about four miles out of town, but being the prairies, it looked as if it was just down the road sort of thing. We said, "Oh, we'll walk," and after four miles at twenty below we got pretty damned cold. We never could get used to the ear muffs and things, but it was almost a courtmartial offence if you got your ears or nose nipped. A lot of fellows did; they just didn't roll their flaps down. But you didn't feel the cold, till after!

Ground Staff, RAF: It was physically cold, but because the sun was shining nobody believed it was cold. Nobody put on a greatcoat or gloves; we just went marching around this damn station. Coming out of England we didn't feel the cold for some reason, and everybody got frostbitten. They finally smartened up and issued orders that people had to wear greatcoats whether the sun was shining or not.

Airframe Engineer, civilian: This English station used to do complete engine changes on six Harvards every morning. By noon they were all finished. They'd roll them out in front of the hangar, three on one side of the ramp and three on the other, facing each other, and as soon as they came back from dinner six guys would get in there and run those things up. Six Harvards, in front of a hangar, roaring at full throttle, and those guys playing around with the governors on the props. Was that ever noisy! But I was really amazed how efficient those guys were, to do six complete engine changes every morning.

Junior Officer, RAF: Somebody was out flying and he came back and said to the squadron commander, "You'd better get all the aeroplanes in; there's a hailstorm coming." And he said, "Ah, we're not going to stop for any hail. Who cares about a little hail?" But the storm started and there were hailstones bigger than golf balls just pouring down, and everybody was absolutely powerless to do anything but just sit in the hangars and watch those Ansons get pulverized.

Afterwards it looked as if somebody had turned a machine-gun on them. I'd never seen such a mess. All the maintenance people were running around with notebooks and clucking away. Literally, if you'd turned a machine-gun on them they would have looked about the same. The storm took about 45 of our aircraft out of service right then.

Student Pilot, RAF: The Royal Canadian Air Force paid us. They flew in the Golden Eagle, as we called him, every two weeks. This aircraft would come, and everybody would say, "Careful! Off the circuit. Get that Golden Eagle down! Don't stop the Golden Eagle!" They came from Calgary where the pay office was, in a Lockheed Lodestar—the paymaster and his assistants and his bags of money. It was all *cash*. "Clear the way!" and everybody would stand back, and in they would come. You went into the Adjutant's office and you called out your last three numbers, and the guy would give you new notes and silver. Oh, it was a great game!

Canada was not the only training ground for British airmen. Before the Japanese attack on Pearl Harbor in December 1941, many of these trainees arrived at No. 31 Personnel Depot in Moncton only to find that they had been selected for training in a warmer climate.

Student Pilot, RAF: From Moncton we were to go on to what was called the ATTS—All Through Training Scheme—where RAF people were posted to flying schools in the United States operated by the US Army Air Corps. The one I was to go to was in Oklahoma, but there were about a dozen of them across the country.

Student Pilot, RAF: We were taken by train to Chicago and there we were marched through the city from one station to another. I don't think the people of Chicago quite knew who we were, because we were in part American and part British uniforms and carrying our kit bags.

We were put on board one of those special trains that ran from Chicago to St. Louis and then on south, and we ate in the regular dining car. The waiters were all really cheesed off with us, because they put out the little plate for tips and we ignored it. We weren't paying for meals, they were provided, and I'm sure the guys that were serving were earning a damn sight more than we were.

Student Pilot, RAF: When we stopped in St. Louis we got out and it was just like somebody had dropped a blanket over the top of you, because we weren't accustomed to the heat. We had been cool on the train—it was air conditioned and we were extremely comfortable—but in St. Louis on the platform it must have been 95 and humid, and in a very short time we all jumped back on the train. That was the first time I realized what an air conditioner was!

OPPOSITE:
Ground Staff, RAF: The summer months came, and the engines in our Ansons started wearing out. They were losing compression, and when we took them apart we found that the cylinders were badly worn—and the pistons, and the rings, everything.

In England all the aerodromes were grass, and aircraft ran up all over the place. You don't have dust, and those aeroplanes had all come out to Canada without air filters. I can remember seeing those aeroplanes taxiing down to take off, one behind the other, with a ruddy great cloud of dust just blowing from one to the next. They were all sucking it up, and we just wore out all the cylinders.

We started robbing from one engine to make up another, but we soon ran out. And then a cargo ship with a whole batch of parts got sunk, so that's when they decided they would go for the Jacobs engine. But in the meantime we improvised air filters by making up a cowl and stuffing it with oily steel wool. It seemed to work pretty well, but by that time the damage had been done and all the engines were pretty well shot anyway.

Student Pilot, RAF: Everything was new. It was all so exciting; it was a new country and you were meeting new people. I was tremendously impressed by the hospitality of the Americans.

We went to Texas, and in those days South Africa would look like Liberty City in comparison with what we found down there. The drinking fountains were marked "Coloured" and "White", and the buses had bare wooden seats at the back for "Coloured Only".

We went to Kaufman County Airport, which was just a small local civilian field that had been converted, and they had put up air conditioned barracks, far superior to the barrack blocks in Moncton. And they had built extra hangars with Link trainer rooms and lecture rooms and parachute facilities. The Commanding Officer was RAF, the Adjutant was RAF, and some of the ground instructors were from both the RAF and the Royal Australian Air Force, but the flying instructors were all American civilian pilots. They must all have belonged to the same organization, because they wore the same kind of half-assed uniform and they had hats that were like the American army hats, but it had a special badge on the front.

Student Pilot, RAF: There were a hundred RAF pilot trainees and there were a hundred American US Army Air Corps trainees, and we were completely intermixed. We were billeted together, and each instructor would generally have two RAF and two American students. They were a little bit older than us, because by '42 people in Britain were being called up younger and younger. We would be nineteen or so, and they would be 23 or 24. Most of them had done college, whereas we had just come out of high school, and most of them were looking to the Air Corps as a career and they were guaranteed to be officers.

Student Pilot, RAF: We used to have a parade every day. The flagpole had two yardarms, and American cadets would go on one side with the American flag and we would go on the other side with the Royal Air Force ensign. We had to pull them up slowly, and all of us had been given firm instructions that both were to hit the top at the same time. Well, there was always competition between the two groups. Each wanted to be the first up, and they would always be edging up, and then they would start going a little faster. And then the sergeant would see what was happening. There was a lot of friendly rivalry.

Student Pilot, RAF: The course content was the Commonwealth Training Plan. The actual manuals, the number of hours you spent on instruments, and the techniques of training were all BCATP. You did the same number of hours on aerobatics, on air-to-air gunnery, on meteorology, aero engines, armaments, and radio, as they did in Canada.

When you got your Wings you were made either a sergeant or a pilot officer, but the Americans *all* became officers, and they got *two* pairs of Wings, an RAF Wings which they wore on their right breast, and they wore American Wings on their left breast. We were all a bit cheesed off. We said, "Why don't *we* get American Wings to wear on our right breast?"

We had our names called out: "You are going to the States for training and you will be issued with American uniforms, but you will continue to wear your RAF hats." So we were issued with American khaki shirts and pants, and brown shoes.

Graduate Pilot, RAF: We'd just got our Wings. We went into Dallas and ended up in a kind of nightclub place—live orchestra—a bunch of instructors and a bunch of us kids with our brand new Wings up. It was one of those places where everything is in darkness except the dance floor and the band, and where the tables have little candles on them. It was all very new to us as Englishmen. We were in the back corner, and suddenly a spotlight comes on and shines over the tables at our party. The guy at the microphone draws the attention of everyone there to these pilots of the Royal Air Force "that are here with a group of our own boys with the US Army Air Corps. We are honoured to have them, and I'll ask them to stand." And the band struck up, "There'll Always be an England" and everybody stood up and applauded. We'd just got our Wings, but we were treated like heroes!

Graduate Pilot, RAF: When you're twenty years old and in a strange country, Denver sounded like an interesting place to go. We were in uniform and we stayed at the YMCA, and they heard us talking funny so they wanted to know where we were from. We told them we were Royal Air Force, but of course the only Royal Air Force that they'd heard of in the States was Battle of Britain.

There was a tap at our door, and there was a businessman standing there. He said, "Excuse me, I'm chairman of the local Rotary Club, and I'd like you to be my guests at our meeting tomorrow." Well, we were young and naive and didn't know what the hell was going on, but we didn't have much money, and "anything for a free dinner". So we went down, and we were *lionized*! "These are Royal Air Force boys . . ." But we did not emphasize the fact that we hadn't fired a shot in anger!

We were in Denver for ten days, and these club members were vying with each other to take us to their homes, to take us for drives, and to meet their daughters. We had to go and meet the ladies who were packing "Bundles for Britain". It was a beautiful experience. To be a hero with no risk attached when you're twenty years old and you've done nothing—it was beautiful!

By the terms of the BCATP agreement, and as members of the British Commonwealth, Australia and New Zealand contributed a proportionate share to the cost of Plan operations and they joined with Canadian and British representatives at meetings of the Air Council. Prior to the war, both "down under" Dominions had training schemes for candidates for the RAF, and it was with political reasoning that they wished to retain some aspect of aircrew training in their own countries. Nonetheless, most advanced training for members of the Royal Australian Air Force and the Royal New Zealand Air Force was carried out within the framework of the BCATP in Canada.

Trainee, RAAF: I always had a great love of ships. I lived in Newcastle, which is right on the east coast of Australia, and after school each day I'd go down to the wharf and I'd watch the boats coming in, and I'd watch them going out, and I'd even ask could I go on the boats. I was just boat crazy, and my mom and dad used to say, "Look, me boy, one of these days you'll be shanghaied and you'll end up in China."

When the war broke out I put my name down for the Australian Navy and I waited . . . and I waited . . . and I heard nothing, and it got a little bit frustrating. Then the thought of the Air Force struck me, and I went down and I applied for the RAAF, and I was accepted.

But I wasn't in the Air Force yet. I was still a civilian attending a school, a reserve sort of thing. You're just waiting until there's a time for you to be called up. I was living at home and I was still

They had two ships that they used to carry troops across the Pacific, the Awatea *and the* Aorangi. *One would come one month and the other would come the next. When we came over to Canada we came more or less as civilians—we had cabins and so forth. We had our Air Force uniforms, but it wasn't like a troopship. It took three weeks and we stopped at Auckland in New Zealand and then came straight across to Vancouver. It wasn't protected; it just travelled on its own, because at that time in the summer of 1941 Japan wasn't in the war.*

IN THE SPRING YOU'LL SEE THE GRASS

They were in uniform without greatcoats, and we thought, "How bitter cold." It went to forty below, but they wore just the Air Force blue and a wedge cap, and just shiny, shiny boots. They were stationed at Rivers, and they said, "Is Canada all this desolate? Is it all like this, cold and snowy and flat? Does it ever get warm here?" And my mother would say, "Wait. In the spring you'll see the grass and the flowers." But they had arrived in the dead of winter.

They used to be very careful of their uniforms, and when they came they'd always hang up the jacket. When they stayed overnight they would say, "Well, I've gotta press my pants." And my mother would say, "Just go and get the iron, dear." But we finally found out that once they undressed they folded their pants flat and would lay them under the mattress for the night.

working, and I used to go to these classes of an evening, or on a Saturday or a Sunday. The Royal Australian Air Force expected fairly high standards for schooling; you had to have reached a certain degree of education, so I brushed up on math and trigonometry and stuff like that.

Trainee, RNZAF: I didn't arrive in Canada till '43 on my way to Europe, a very impressionable eighteen year old. I came from those little islands of New Zealand in the south Pacific and the first thing that struck me was the vastness of North America. We landed in San Francisco and travelled by train north to Vancouver, and then crossed from Vancouver to Macdonald, Manitoba.

Trainee, RAAF: After we'd been in Vancouver for a day we got on the train and most of us were struck with stomach disorders. Some had it so bad that they would just sit in the carriages in their underpants because they figured they didn't have the time to keep putting their uniform on and then take it off and get back to the toilet again. There was always a line-up at each toilet on the carriage, and they used to go 'round to the other carriages. It was a real troop train; there were no passengers on it at all.

Ground Staff, industry: When the Aussies first came all the girls had to have an Aussie boyfriend. And then the New Zealanders came and *they* were the big thing. I met one that ran around with my chum's sister. She'd bring him home and we'd be sitting at the table yakking over a cup of tea, and it was more or less an oddity just listening to him talk, and his expressions. We were asking him about what he did in Australia, and I just about fell off my chair when he said, "Oh, I had a good screw before I left." What a thing to say! His girlfriend sitting there, and her mother, and the rest of us sitting around. I didn't know *then* that it meant a job.

Trainee, RAAF: One of the things that they stressed to us when we left Australia was, "You're going overseas and you are ambassadors of your own country. Let's act accordingly." We did have black sheep in some of the groups, but nobody sort of spoiled their copybook when they went to anybody's home. And after a period of time you almost felt that you were one of the family; they accepted you so well and their homes were always open to you. They usually had a son overseas, or a son in the Navy, or a son somewhere else in Canada, and they were hoping that their son would be looked after the way they were looking after us.

Canadian Host: They came to us at Christmas. The Red Cross had asked for people to open their homes to airmen, and it was Christmas Eve and everything was shut, but they turned up with a box of chocolates and a copy of Hilton's *Random Harvest*. It was rather staid at mealtime—they didn't know us and we didn't know them—but my mother could draw them out and ask about places at home and that sort of thing. My brother was overseas, and she was a very sentimental person and she used to say, "I wish someone would look after my son, so I'm going to see what I can do for somebody else's boys."

They slept in my brother's room, and after they had been there the once they would come whenever they had the chance. They came weekends and they just fit right in. They were no longer guests, and we just went about our daily business.

Trainee, RAAF: We had what they called bum freezers; they were little short coats and only came to about here. We got to Canada and they were pretty cold and the Air Force then issued us with the regular Canadian greatcoats.

I used to go through Assiniboine Park in Winnipeg because we used to get a streetcar that took us up Portage Avenue and we'd get off on the north side of the park and walk through to the wireless school on the south side. It used to be beautiful walking through there of an evening, and I was always fascinated with the snow. I got to know afterwards that it wasn't always fascinating.

Canadian Host: One time there were a number of young men they thought they would like to bring with them. Mother said, "Certainly, bring them along." So we scraped around to get a few girls to make a party, but it was just very simple entertainment. My parents didn't drink, so there was no liquor allowed, but there was plenty of food and coffee and that sort of thing, and we all cleared out of the dining room and they danced a bit. We had a piano, and they would sing—mostly the old kind of songs that people do sing around a piano.

One of them, Tony, fell head over heels for my sister. We'd be working in the kitchen and Wally, the other one, would come out with his hands in his pockets, kinda mooching around. And if you went into the living room Tony would be sitting with his arm around her, and Wally felt like the fifth wheel, you know. It never came to anything, just one of those things—the attraction that's there. They were young, and I'm sure they were lonely.

Trainee, RAAF: I met my wife through a blind date. We started going together, and her folks invited me out to their place for dinner one day, and it just seemed to go from there. I used to go there every weekend.

People were very good to us. They sort of opened their arms to us and brought us in, and they didn't really know who we were.

Trainee, RNZAF: I had no difficulties in assimilating with Canada, the people, the climate, and that sort of thing. We just hit it off nicely with the Canadian civilians and the boys in the service. The camaraderie and the friendships developed very quickly. You always had a certain competitiveness; you didn't stand up and wave your country's flag around, but there was always a personal pride and you were prepared to defend it. It was a nice understanding, a good relationship.

Canadian Host: We had English boys who were fascinated by the weather. They found it very bitter, but we told them, "You can have fun in Manitoba in the winter," and we took them sliding. It was just a little sled about four feet long with a flat body—just a little wooden sled—and a toboggan.

We lived near a river and there was an embankment, and all the local kids used to go there. They only had uniforms and they had to keep them decent, so we scratched around looking for clothes. Vic could wear my brother's, but David was much taller and he was a little stouter, and my father was a rather stout but short man, so he had to wear Dad's clothes. We had loads of mitts and scarves and stuff, but it was these pants of Dad's. David got them on, but he had to belt them in, and then of course they came at half-mast up his legs. But they thought this was the greatest thing. They'd never dreamed of sliding down a hill on a kid's sleigh. This was totally new to them, and probably not the dignified thing to do, but they thought it was great.

Wife, RCAF Instructor: The wives of some of the instructors would go out to the station and help at the canteens. Or we'd sew, or mend socks and things for the airmen. And when the Australians graduated we went out and sewed on all their Wings. They appreciated that, because there was really no way they could do it themselves.

Wife, RCAF Instructor: The night before he graduated one of them came to the front door and said, "You know, Mom, I wanted to say goodbye

Trainee, RNZAF: I remember playful skating competitions and crawling down the ice rink on our hands and knees, pushing a stick in front of us and dragging our skates after us. Most of us had never seen skates in our lives and we tried it because we had that inbred want to try it. Australians and New Zealanders would challenge each other to an ice hockey game; none of us could skate worth a damn and it would all end up in a sort of rugby free-for-all.

to you," and he brought me his flying vest. It was sleeveless and all sheepskin-lined, and he said, "Here, Mom. I want you to wear this when you go outside to hang out the clothes, so you won't catch cold." And he said, "You can think of me when you wear it."

Trainee, RAF: They had a wonderful hospitality set-up behind Eaton's in Winnipeg. They would find out what your interests were and what your job was back home, and then they'd send you to various homes so that you'd have like interests.

They had a dance floor there and the inevitable juke box. I remember they were always playing "Paper Doll", and you didn't have to stand around very long before you were invited to a dance, or a refreshment or a meal, by very charming girls.

I made such good friends I went back to the same people whenever I possibly could. One particular fellow took me to the various clubs that he belonged to and we'd go to dinner at them. And I remember going up to Winnipeg Beach on the old train, on a picnic. An old taxi driver took me up there.

Trainee, RAAF: I was impressed by the conveniences like washing machines and other things that we hadn't been used to in Australia. And the big refrigerators—a lot of the stuff seemed to be *big*, and there seemed to be so many appliances that made housekeeping easier for a woman than was the case at home. Australia wasn't backward, because we did have lots of things there, but it just seemed that people had *more* over in Canada.

Trainee, RNZAF: Only once in my life have I ever been homesick—in the first year I was in Canada when I was on duty New Year's Eve. During the war years not everybody got New Year's off; someone had to stay and sort of guard the fort, so it happened to be my job to be part of the Commanding Officer's reception party for station personnel. This went on until the Grand Old Midnight, and it wasn't until everybody stood up and held hands and began to sing "Auld Lang Syne" that I felt my first pangs of homesickness.

During the years of the BCATP Canada was visited by airmen in training from most parts of the British Commonwealth and from many other countries that had been overrun by the enemy. Generally, these trainees had joined the RAF or the RCAF, but even at the time of Dunkirk, members of the Royal Norwegian Air Force who had escaped the occupation of their own country were a well-organized group. They then joined the Plan as a separate, but integrated unit.

Officer, RNAF: The Norwegian government had established an organization in London and they started gathering Norwegian citizens that were floating around. Finally we assembled enough to start an air force, and we were then looking around for a training place.

We got a message that we were going to an island in Lake Ontario, and this sounded a little funny: "We're going to be put out on an island ... in Canada?" We didn't think that this was a very good idea, but when we got there we found that there was just a little channel to get across to Toronto, and that was pretty nice.

Ground Staff, RNAF: To get over to Canada was quite a chore, because we couldn't get on any of

SEWERS DIDN'T WORK

Ground Staff, RNAF: When the first Norwegians arrived in Canada it was getting late in the summer and we were looking around in Toronto for a place to build a camp. Lakeside Homes was a summer place for sick children on Toronto Island; it was run by one of the hospitals, but it had been sitting idle for several years, and we were able to get the use of this. When we moved in the sewers didn't work and there wasn't any heating, but there were gas lines, so we got some of the gas working and we got some heat in the buildings and stayed there until November.

We got the use of the property behind Tip Top Tailors near the Toronto Ball Park. Between there and the channel to the Island Airport was a piece of land nobody used, and we built our camp there on the mainland. We moved in about the beginning of November in 1940.

The aeroplanes that we had were Cornells; we called them Fairchilds. A lot of them were paid for by rich Norwegian North Americans who had emigrated and

Airman, RNAF: *In the beginning some of us had Norwegian uniforms, but some others didn't have any at all. They took the basic colours of the Canadian Air Force, but we used the Norwegian insignia and rank distinctions, and the Norwegian flag on the shoulder. That was about the only difference.*

Ground Staff, RNAF: *The existing Island Airport in Toronto was run by the Harbour Commission, but there wasn't any flying hardly at all. We were able to get the use of this, but we had to get across every day using a ferry.*

THERE WAS NO HEATING

done well financially. We had about fifty at the end. Our markings were a blue fuselage and the wings yellow, and the Norwegian colours on the rudder, and red, white, and blue stripes across the wing, top and bottom.

Student Pilot, RNAF: The instruction was started out in Norwegian, but then to serve the purpose of the coming proficiency in the English language, the major part of the instruction was done in English. The next step was the advanced training at Camp Borden, and that was all in English; there was no exception. Only when we had trouble adapting ourselves to the aviation terminology in English were we allowed to switch over to Norwegian. Most of the instructors that we had there were either Norwegian boys that had been trained very early in England, and also in Canada, or they were pre-war instructors from Norway.

Later on traffic at the island became too heavy, and we had to move up north of Toronto to Muskoka.

Airman, RNAF: *When we were building our camp on the mainland, we were still living on the island, so we had to go back and forth every day using a little boat.*

Officer, RNAF: *At Muskoka we had a hangar and barracks that were built by the Norwegian Air Force. The Officers' Mess and the dining rooms and kitchen were built out of logs by a group of Finns, Norwegians, and Swedes who were sort of gathered up from around that part of Ontario.*

the British transports, but we managed to get hold of two ships that had been used in Norwegian coastal traffic. They were fifteen hundred tons each, which was pretty small for the Atlantic. They were coal-fired steam engines, and they didn't go very fast—our maximum speed was eight knots, and the convoy that we were assigned to went at nine knots. So we were left behind quite soon after we left Glasgow, and came across the Atlantic by ourselves.

We had some engineering types and we managed to get former Norwegian Naval Air Force people as deck officers, who were qualified as mates and that sort of thing, and we found two skippers in Scotland that had come over from Norway. We also had several future officers and a future Chief of Staff of the Norwegian Air Force on duty as stokers in the coal room. So we manned these ships ourselves and there were about fifty or sixty people on each. So there were about 120 of us that came to Canada in that first group that arrived in August of 1940.

Officer, RNAF: People were sent out to Canadian stations to find out how they did things, and we then programmed our training in the same system as the Canadian Air Force was organized.

Student Pilot, RNAF: In mathematics I had a Norwegian Canadian from Vancouver and I also had in geometry a lady teacher from the United States. They spoke perfect English, but also could be understood in Norwegian, then as we advanced in the handling of the English language, we graduated into sections where the teaching was done in English.

Officer, RNAF: We came to Canada with a number of pilots that were already almost fully trained pilots. They were Norwegian Air Force people and they went on the P-36 training program, and after about two months they went overseas as full-fledged fighter pilots. By that time we had several instructors qualified for the P-36, and they did the job of checking people out. The students that started their flying from scratch in Canada then went on from the Elementary School at the island to service training in the Canadian Air Force and were part of the Commonwealth Training Plan.

The Norwegian trainees continued to arrive. They were the only group, other than the British, that had a separate flying training operation under the control of the BCATP. Within a very short time they had become welcome visitors to Canada, most particularly to the citizens of Toronto.

Officer, RNAF: I was technical, so I was sent to St. Thomas, where they had a big training school, and I went through every important part of that in a matter of a month. When I came back to Little Norway I was then put as an instructor in the mechanics' school.

Student Pilot, RNAF: Every fortnight we were allowed to go anywhere we wanted, provided we had either some sponsors or we were capable of looking after ourselves. If two or three chaps went together they made sure that one of them was proficient in English.

The units were Norwegian in the elementary flying, but we did the same number of hours as the Canadian students and we took the same courses in ground school as the Canadians did. It was all taught in Norwegian at first, and it was all taught by Norwegians. It consisted of ground training for air crew and another school for ground personnel. And then the elementary flying was carried out using the Fairchilds.

We were fed propaganda right from the beginning: "You do this for your freedom; you do it for your flag and your King, and your country." We probably did feel that way, but I think for most of us this was an adventurous thing. You went into it because you liked the excitement, and the patriotism came second.

Student Pilot, RNAF: You could have a multitude of passes provided you could fend for yourself and ask directions if you were lost. One way that I learned to speak English was watching the cinema, starting at six o'clock in the evening and finishing at eleven at night. Seeing the same picture over and over, and lip reading to see how people pronounced the "Ws"—we pronounced them "V"—you have no idea how fast you pick up a foreign language by doing that. It was mostly cowboy films, and of course you still had Clark Gable, and there was *Gone With the Wind*. I must have seen *Gone With The Wind*, without a lie, ten times.

Ground Staff, RNAF: The first Christmas we were in Toronto we had invitations from private families for Christmas dinner and we found we didn't have enough people to go to all the places. I was at a Christmas party, and presents came in boxes—boxes *full* of them. That first Christmas in Toronto the boys got more presents than they had ever had before. They would get a pair of socks, and others would get mitts, and this sort of thing . . . but very often they'd reach into a pair of socks and find a fifty dollar bill!

Officer, RNAF: Traffic on the airport increased, and when the Americans got into the war there was an awful lot of traffic between Toronto and American stations. There were military aeroplanes and civilian aeroplanes that came in, and they became a bit of a problem. Besides that, there was a problem with the short runways, and so we got Muskoka Airport to use, and we moved up there in '42.

Student Pilot, RNAF: As Norwegians we were really motivated. We hated the Germans intensely because we were invaded in 1940 and because news had filtered through that our families back home were mistreated, and a lot of our parents were already in concentration camps. So we had a different outlook from the beginning on the purpose of becoming a pilot. I wanted to go back and kill as many Germans as I possibly could. It was personal.

Student Pilot, RNAF: We had a German prison camp not far from Muskoka, and this instructor with a student on a night flight—they had pasted tape over the markings on the tail—he flew low over the cook's hut inside the German camp and dropped bricks through the roof and made a hell of a mess. There were six Cornells up night flying, but in spite of the tape he was identified. He had hoped this would send him to England or a posting to twins, but instead, the Commanding Officer said, "No, you're going to spend another year instructing!" But of course he was the hero of the camp, because we still didn't carry much love for the Germans.

Student Pilot, RNAF: For my service training I went to Hagersville, which was set up for Ansons. It was really a low performance aircraft compared to the Cornell, so we felt that this was a little set-back in our training. We had some Norwegian instructors at Hagersville—that was part of the agreement with Canada, that we would provide some instructors—but most of the instructors were Canadians, and I flew with Canadian instructors more than I flew with Norwegian ones.

Student Pilot, RNAF: There wasn't much Norwegian in Camp Borden, except that we had an officer whose shoulder you could cry on—a liaison officer more or less. Two of the flights were commanded by Norwegian officers, and we had on the same flights two or three Norwegian instructors, but the Canadians were in the hundreds. In the ground school we had all Canadian instruction. You're still not really proficient in English to write well, and they thought they were going to help the Norwegians by making multiple choice questions in some of the examinations. But that confused us even worse, because in a multiple choice the shade of one sentence can screw up the whole deal. You can be on the right track, but think wrong. I still had problems with English, even though at Camp Borden we had special classes for teaching it.

THE GOOD TIMES

The Harvard had the power and the neatness, and yet it was a very dangerous machine. It would flick very suddenly. If you pulled it round a loop too quickly it would flick stall, and in that way it was a very tricky machine to fly. It didn't give you any warning, it would drop a wing so quickly. A lot of people came to grief with them.

IT WAS A REAL THRILL

Student Pilot, RCAF: For three years before I went off to the Air Force I was a boy at Bishop's College School just outside Lennoxville. When I was stationed at Windsor Mills I paid a visit to my old alma mater. It was only a few miles away, and I followed the St. Francis River and made a few low passes and pulled up into a slow roll and then thought, "Well, maybe I'd better let well enough alone," and I decided to get on my way back to Windsor Mills before I got into any sort of trouble. But that was a real thrill for me!

Trainee, RNZAF: When we were on our way to Macdonald our train stopped at Golden in British Columbia, and we were told we were near the Great Divide. We got out and had snow fights and the like, just grabbing a whole bunch of snow in our hands and wrestling all over the place. Snow was something new for most of us.

Airwoman, RCAF: They'd never seen ice before; our men taught them to skate. They could barely stand up, but they'd challenge our guys to a hockey game. Of course, the WDs all fell in love with their beautiful accents.

Ground Staff, RAF: I met a very pretty lady and asked her to marry me. I had to get permission from our CO. You weren't allowed to wear civilian clothes, and he said, "Is that your best uniform? Let me have a look at it." There was a split in the tunic, and he said, "Before you get married, you'd better get yourself a new uniform." But he gave me permission. We were at the station church at eight o'clock on the Saturday night, and when I turned around to walk away from the altar, to my amazement, the CO and the Wing Commander who was his aide were both standing there. I felt so proud!

WD, RCAF: *Joining up just seemed the thing to do at the time. I had been knitting socks and everything else, but I felt that it would be sort of more worthwhile—and that I'd be replacing a man. That's what we all felt.*

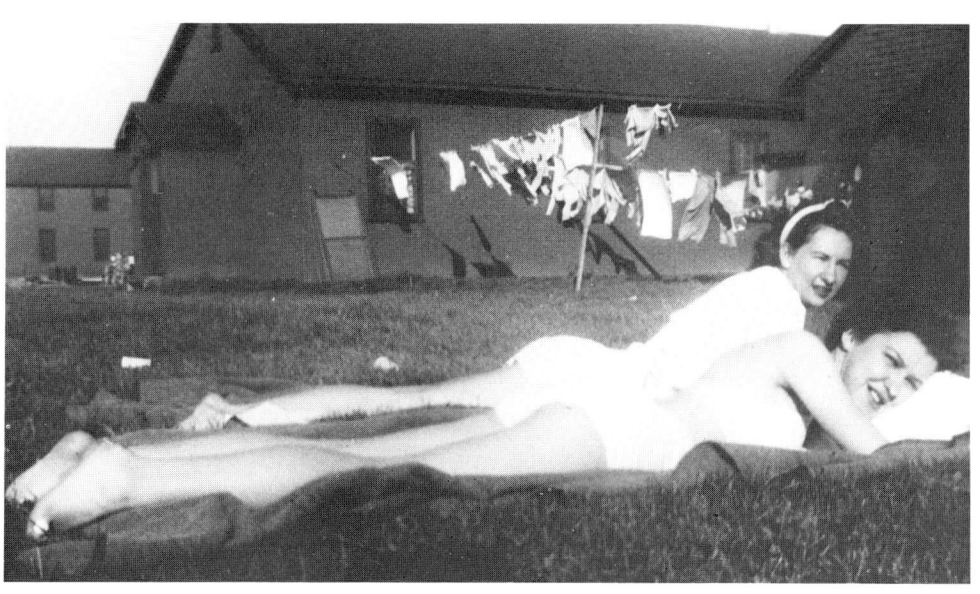

Student Navigator, RCAF: *One of the stations I was on seemed to be about twelve miles from the nearest village. It was away out in the sticks, and we used to spend our time off walking the railway tracks because the roads were too muddy. But it wasn't all that bad, because there were quite a number of WDs on the station, so we were never short of company.*

Student Pilot, RNAF: *We had a fantastic reception from the Canadians and there weren't enough boys to fill the invitations for weekends, and so we had a choice more or less of where we wanted to go. This was summertime '43, so my first impression of Canada was nothing but peaceful living out in the country and swimming on lakes. We were rather overwhelmed by the hospitality that was shown us.*

We were close to Grand Bend. It was a good place to go for dances and swimming, and on your free weekend there were cabins that you could rent. It was a little resort just near Centralia, and there used to be a busload of us go over every weekend, and we'd party and swim and stuff.

FAR LEFT:
With about twelve bucks in my pocket I went to New York for three weeks and had a great time. It was wonderful to go along in that RCAF uniform. Of course, they were warm, and down there the humidity was a bit of a problem—trying to be properly dressed in those conditions with a winter uniform on. But we had lots of fun.

OPPOSITE:
A couple of us girls hitch-hiked down to Pugwash in Nova Scotia. If you stayed in uniform people would pick you up, but we got outside Dorchester and somebody stopped and said, "Get in the car real quick; we're not supposed to pick up people because there's a penitentiary here." But because we were in uniform, they took a chance.

I remember in Quebec seeing little carts being pulled by dogs and carrying cans of milk. People would have them beside their houses and we could rent them for a buck or two and go for a bit of a ride.

Student Pilot, RNAF: *We had a Commanding Officer who was very well liked, and one of the statements he made, and which he may have regretted later on, was, "The only way for you kids to learn the language is to go and find yourself a girlfriend," which quite a few of us did!*

Student Navigator, RAF: *We used to have horses at Rivers. I forget what we had done, but I remember for the whole week before our Wings parade, about four of us had to shovel horse shit out of the stables.*

Wife, RCAF Instructor: *We had a wedding; one of the WDs and one of the airmen got married. They were just going to get married at the station and then leave from there, but we said, "No, you'll have a reception here." There were 25 or 30 people, and I made a cake and put a little aeroplane on top. Our place was a meeting place. You just left your rank outside. One night we had a Wing Commander sitting at one end of the chesterfield and an AC2 at the other. And we had WDs. Some of the WDs had never been away from home before, and they always called me Mother. If anything went wrong they came to our place, and I felt if I could help them in any way, I was glad to do it.*

Airman, RCAF: *There wasn't much money around, but we'd make the best with what we had. Then our leave was usually so short that we couldn't go very far anyway.*

Officer, RAF: *We were way out in the sticks. Winnipeg was about the closest spot, but we didn't get there too often. Mostly we made our own fun and invited the locals into the mess.*

Sometimes the boys could bring their girlfriends, but the girls couldn't come in a mob—they had to be escorted. If you were bringing a civvy you had to have the CO's permission and you had to report them in and out, and leave the name with the SP at the gate. They wouldn't stand for any gate crashers.

There was a standing invitation everywhere you went, and people would say, "Come and see us," or "Would you like to do this?" We had no difficulty in keeping busy on our time off, and our hardest job was in reciprocating in some way. I was the guest of various families at different times, and I developed a real affection for them.

Instructor, RCAF: *The Anson was like a big truck, and you couldn't throw it around—you couldn't manoeuvre it—and I loved to be able to manoeuvre an aeroplane and make it do all sorts of things. But an Anson is not intended to be thrown around quite like a Harvard; you don't aerobat it. But it's an easy aeroplane to fly—you can't get into much trouble in an Anson.*

Flying really consists of two things; the thrill of flying is the thrill of taking an aircraft off, and landing that aircraft on the ground. In between may not be the most exciting thing in the world.

Every pilot, against all instructions, could not resist the temptation to low fly. It was the greatest sensation of flying, the closer the ground the greater the sensation of speed. 100 mph at ten feet is much more exhilarating than 100 mph at 1,000 feet. So we all tried that. Some of us got caught and were disciplined. Some of the fellows were washed out, some of the fellows wrote off aircraft, and some of them killed themselves.

Being a BCATP staff pilot was generally considered boring and mundane; driving a Fairey Battle for gunnery students, an Anson for aspiring young bombaimers or navigators, or a Norseman, Moth or Fleet Fort for wireless students was hardly the dream of the young pilot with visions of flying Spitfires. But there was one branch of Canada's wartime Air Force that resembled the growing commercial airlines of the late Thirties. The staff pilots there were not only akin to the dashing young airline pilots of the day, but also faced the personal challenge of carrying Very Important Persons, a task requiring varying degrees of discretion.

Staff Pilot, RCAF: I was posted to a squadron at Rockcliffe, the Communications Squadron they called it. We were flying brass out of RCAF headquarters, just hustling them around the country; we'd go coast to coast. There were many times that we carried Billy Bishop—the whole station would be on parade for his entrance and inspection.

Senior Officer, RCAF: Billy Bishop was back on the job as Master Recruiter, and he went the length and breadth of Canada making speeches and stirring people up. He did an extraordinary job of creating interest in the whole Air Force operation.

Officer, RCAF: Bishop was a glamour boy, and his job was making speeches and getting us recruits. They called him Director of Recruiting, but we had a Director of Manning who actually did the recruiting and we had this Director of Recruiting whose job was strictly publicity. Bishop was our PR man.

Senior Officer, RCAF: He was very much intrigued by having been asked to return, but he had that quality to be able to talk to groups and individuals and to be inspirational. It was really a PR selling job. He had a great record in the First War and he certainly was able to induce lots of young men to zero in on the Air Force as their aim. Of course, all this time the Army was after people, and the Navy also, but neither of those had a Billy Bishop.

Student Pilot, RCAF: His polished image during the war must have been very important—just bouncing around as a figurehead. He certainly was a hero from the First War; just how much he really deserved to be the kind of hero he was in the Second War I don't know, but the country needed somebody and he was there. He did the best possible with what he had.

6

Billy Bishop builds an Air Force
Hollywood hits the BCATP
The WDs and their growing pains
Three Rivers to Calgary on $14
Thumbing it
You're on a 48 and you're AWOL
The true colours of Quebec
Anything but conscription
Spare parts were non-existent
All I had was my little tool box
Paper aeroplanes?

THE FRIENDS & THE FAITHFUL

Staff Pilot, RCAF: I was posted to a squadron at Rockcliffe, the Communications Squadron they called it. We were flying Brass out of RCAF HQ, just hustling them around the country; we'd go coast to coast stopping at all the training stations. One of our aircraft was an original ten-passenger Trans-Canada Air Lines Lockheed 10; they gave us those when they graduated to Lodestars.

Student Pilot, RCAF: There was no question about it, he was very inspirational, and most of us, if not all of us, aspired to follow in his footsteps. He was an Air Marshal at the time, which was a rather unheard of rank, and as a young kid I admired him very much; I was awed by the man and his rank.

Senior Officer, RCAF: He hadn't achieved the rank of Air Marshal in the ordinary way. He was of course retired when the war started, but when this job was created for him, and he was to be the sort of key figure trying to interest the young people and to get recruits, he was given the rank of Air Marshal, which he enjoyed. He certainly did everything that one could ask and he worked very hard—on the go all the time.

Staff Pilot, RCAF: One morning I took him to Montreal. I dropped him off at Dorval, and he told me to contact him the next morning "when you come back to pick me up and we'll be going back to Rockcliffe." I got there punctually and I phoned him and I said I was "here and I'm ready." He said, "Oh, I'm not going back today. You go on back to Rockcliffe and I'll call you again when I'm ready." So it was about three days before I got him back, and that was certainly dollars wasted. There was an awful lot of that in that particular area.

Of course, the poor man was a heavy drinker, and he was right out of it many times at Rockcliffe. We couldn't load him at the ramp; we'd have to meet him at the button on the runway. They brought him right out there in his limousine, and I simply did what I was told.

Senior Officer, RCAF: He did a remarkable job, and someone else who did a remarkable job was his personal assistant who helped to organize and look after him, because he had a few rather eccentric mannerisms and he had to be helped a lot. One of his problems was that he enjoyed a convivial evening and at times he went overboard a bit, and I personally, when I was a junior officer at headquarters, was delegated to see that he was able to get home in one piece.

Staff Pilot, RCAF: The young eighteen year olds who were coming into the Air Force were actually impressed by his past experience as a First World War air ace—that meant something to these young people. And he was the kind of fellow that you had to look up to, both in his civilian career as well as his past experience in the Air Force.

Ground Staff, civilian: We didn't have a big pool of people in Canada to form an adequate supply of instructors for this British Commonwealth Plan. Aviation was much more popular and more available to the ordinary person in the States than it was here, and there were young fellows down there who were taking flying lessons on Pipers and Taylors and things like that. We had people that went down through the States looking for pilots that had some flying experience to make instructors out of them. We had fellows that came up here that ranged all the way from nineteen years old to over forty. They recruited these people, but they weren't very fussy about personal references.

Instructor, RCAF: There were an awful lot of Americans that came up here and instructed in the early part of the war, mostly in 1940. A lot of them sensed that sooner or later the USA was going to be dragged into the war and what they had in mind was that by getting a jump on things, in the long term they would get higher rankings in their own Air Force when the day came that they would be required. And I guess that's exactly the way it worked out.

Instructor, RCAF: *The film people took over at No 2 SFTS for a few days. We had a Wings ceremony and a big parade through Ottawa, and of course a long propaganda speech by Billy Bishop.*

Ground Staff, RCAF: *James Cagney was the main star of the show, and of course Bishop was around a lot. Then there was all kinds of flying with Harvards and stuff, but mostly I remember just waiting around.*

THEY'D NEVER HEARD OF US TILL THEN

Instructor, RCAF: When I was at Trenton in the summer of '41 Warner Brothers were making a picture called *Captains of the Clouds* with James Cagney, and we got into some formation flying that was kind of thrilling. The film was about bush pilots and instructors and all the things that our Commonwealth Training Plan was involved with, so they had big formations of Ansons and Harvards. All the formation flying that we had done up till then had been just two aeroplanes—one on one sort of thing—but here we were in formations of 12, 15, and 18, which was really much more fun.

Junior Officer, RCAF: There was hush-hush recruiting in the United States. We got quite a few people from there; in fact, they flocked up. Bishop did some speaking down there, but there were certain political restrictions.

Senior Officer, RCAF: I don't think I would call it Air Marshal Bishop's scheme to recruit people in the States, although he was very active in that field. It was an Air Council arrangement, where a Group Captain, a resident of the US, was the key man south of the border in heading up the recruiting machine, or whatever it was. But Bishop did a lot of good will in attracting recruits from down there.

Ground Staff, RCAF: *They shot some scenes about bush flying at a lake somewhere. There were a lot of old aeroplanes on floats and one of them was the first Norseman that had been built back about '36.*

Official, Canadian Government: We had to go carefully, whether we were looking for people to be trainees in the Plan, or to be trainers. I'm not sure that it was legal under American law to train people to fight in the Canadian services, but it may have been legal to have them training our people. If they came on their own, we could take them, but whether we could go down there and have someone like Billy Bishop actually making contracts with them on behalf of the Canadian government, I don't know.

Ground Staff, civilian: *Cagney played the part of a Canadian bush pilot who offered his services to fight only to become an instructor. He did a lot of fooling around with aeroplanes, but in the end he became a hero and died winning the war.*

Officer, RCAF: *There must have been a lot of co-operation between the BCATP people and the film producers.* Captains of the Clouds *opened on Broadway in New York with all sorts of praise for the RCAF; I guess they'd never heard of us till then.*

PR Officer, RCAF: *Half-way along the road some of the public were getting just a small bit cynical about Bishop: "Oh, there's too much Billy Bishop." But it was not Billy's fault; we were asking him to do as much as he could, and he was damned good—he wasn't trying to show off. You could send him into a manning depot or a flying school and he'd talk to them, and their eyes would light up—they were thrilled.*

Ground Staff, civilian: We had a group of people that had been Reserve pilots in the US Navy—the weekend types—and these people came up. I remember in particular we had one who was pushing fifty—just a whale of a nice guy.

Instructor, RCAF: There must have been some recruiting for the Americans to hear about us. In the three special advanced flying courses at Trenton and Borden in 1940, there were about sixty men, and roughly three-quarters of them were Americans.

Ground Staff, civilian: After Pearl Harbor there was a repat party that came through to get some of those people to go back to the States. We had one Yank flying instructor and they asked him if he'd like to join. He said, "Oh, I'd like to join the Air Corps." But, "Oh, gosh," they had no openings in the Air Corps at the moment but lots of them in the Navy: would he like to join the Navy? "No, Sirree," he said, "No way I'm gonna join the Navy. I take off from the carrier and I go out, and I fly the mission. I come back and I'm all shot up, and I'm bleedin', and I'm tryin' to find the carrier, and there's nothin' there but an oil slick. No, Sirree, no damn Navy for me!"

The decision was made that the RCAF should recruit women. There were women in other Allied services, and there were many trades in the RCAF in which women could perform the same tasks as men—in some cases they were better qualified. Nonetheless, there was a certain amount of resentment to the WDs (Women's Division personnel), and some of it was found in unexpected places.

Airman, RCAF: We were told that the WDs were coming, and we knew ladies in town that were very upset that the WDs were going to come to the same station and live in the same barracks as the men! I wasn't concerned about that and I never heard the fellows talking about it, and of course they weren't in the same barracks, but there was talk in town among the civilians.

Airwoman, RCAF: We weren't too welcome at first. There were men lined up on both sides of the road just out of curiosity: "What did these girls in uniform look like?" But they felt that we were stepping into a man's world, and we were not quite as welcome as we would have liked to have been. We weren't allowed to mix with any of the officers.

In the beginning the men disliked us very much. We were the first airwomen on the station, and they resented it. They played a lot of nasty tricks. I was driving into Dunnville to get the CO and there was a peculiar smell in the car. It was in the winter and cold, and he started sniffing, and then I started sniffing, and I thought, "What in the world is it?" When I got back I found that one of the airmen had put a hunk of Limburger cheese on the manifold.

You always had to sign for things. They'd hand you a screwdriver or something through the wicket and they'd have an electric wire attached to it, so you got a shock. It took a while for them to get accustomed to the fact that there were going to be women in the service.

I was a farm girl milking cows, but as soon as I was eighteen I joined up. It was quite a change going from a farm where we had quite an isolated life, to moving into a barracks block with 25 or 30 women sharing one room. I was very homesick for a while, and some of those nights you could hear the girls sobbing all through the barracks, especially when we got our needles and everybody was sick from the effects. It was my first time away from home.

We took aptitude tests when we signed up, and they told us what trades we could go into. There was just something about wireless I thought would be kind of interesting.

I went to Montreal and took my course as a wireless operator ground. We had to learn procedure and the Morse code, how to operate the machine to transmit or to receive, and quite a bit of theory. Then I was posted to Boundary Bay, BC. They had a lot of trainees from England there—from the RAF—and it was strictly training that they did. I was in a part of the control tower, and we were in contact with the planes once they took off.

I had to be a court stenographer at Courts of Inquiry. In one case a couple of pilots had crashed and I had to take the notes. I'd work all day in the sessions and then they'd lock me in a room and I'd spend the night transcribing the notes. Then I would get a 24-hour pass, and go back at it again. I couldn't talk to anybody and I would be sleeping when everybody else was at work, then I'd be working when everybody else was sleeping.

Ottawa wasn't a very nice place for airwomen. The civilian women didn't like the girls in uniform because they thought we were taking their boys from them, and in the stores they'd just as soon walk over you as look at you. There was only one place we got any service, the rest of the places they just ignored us. Yet those same girls were permitted to come into *our* canteens and buy Kleenex, and jam, and stuff that was rationed during the war. And that used to embitter us a wee bit, because they were so terribly rude to us in the stores.

My female Admin Officer said they were looking for service policewomen, so I went off to Trenton for training by the Mounties.

We walked the beat in Ottawa. We wore red flashes and arm bands and we lived off the station in a private home. We weren't a very popular lot. We used to walk up Sparks Street to Bank, and down Rideau past the Union train station. The first duty I did was in the summer along the Driveway, along the canal. We had to patrol the park there. This one night when I went out there were just mobs; Army, Air Force, and Navy types all along this path. Every place I went there would be somebody, girls too. They were having a big time. I couldn't have controlled any discipline if I'd tried, because they were really having themselves a ball.

My trade was called fabric worker, and I doped roundels and markings on the aeroplanes. We were taught about masking and to be very precise about the markings underneath the wings. Working in the dope shop was considered fairly dangerous because you could succumb to the fumes, particularly if you were a person that didn't smell them. You had to get lots of fresh air and we had to take three bicarbonate of soda tablets every day and drink a quart of milk.

It was great to be in the Air Force. You're helping out; you're helping your country and there's something to be proud of. You're doing something worthwhile instead of just fooling around.

I was proud of my uniform—I really was—and I had it pressed, and the buttons shining, and the shoes shining all the time. You sort of lift your head up high; I really liked it. Every time when I came home my father would say, "Come on downtown with me." And he would be proud of me and he would say to people, "This is my daughter. She's just home on leave."

My father was of Dutch extraction. He came to Canada from Holland in 1907, and he fell in love with the country and he became a naturalized citizen immediately. He was very patriotic, and I think the nicest thing I ever did for him was to join the Air Force. He was proud of being a Canadian, but he was also very proud of me.

*****

An important part of life for everyone who was involved with the BCATP was leave, whether it was a 24 or 48-hour weekend pass or for fourteen or thirty days, it was a welcome break from the routine of training. Canada was emerging from the Great Depression, and although the Air Force trainees were paid only limited wages, living and travel were usually at government expense.

We went on travel warrants. It cost us $14 to go from Three Rivers in Quebec to Calgary in Alberta and back again, with meals. They recognized the fact it wasn't our fault we were in Quebec instead of training right at our own back door. You slept on a seat on the train unless you wanted to pay extra and get a berth. But they had damn good meals on those trains, wonderful meals. We had good accommodation and they treated us right.

You could go to Windsor and walk across the bridge and the Americans would have fifteen or twenty cars lined up wanting to know where you were going. If you said, "Nowhere," then you went to their house for the weekend. They weren't in the war at that time, and they would be at the Detroit end of the bridge just waiting for service people from Canada, particularly if you had a white flash in your hat.

We were always a novelty, and they thought anybody with a white flash was something special. We'd normally go in twos and threes, and we'd always be picked up within five minutes.

I HAD TO SNEAK MY BIRTH CERTIFICATE

Junior Officer, RCAF: We had the WDs, airwomen—there must have been a hundred or so on the station. Some of them were parachute riggers, but they were also in the Motor Transport Section and on the kitchen staff as cooks. And we had one or two that were airframe mechanics.

Airwoman: I had to sneak my birth certificate to send it to Ottawa to prove my age. My mom and dad didn't want me to join the Air Force because they thought they would never see me again, but I wanted to work on aeroplanes, and this is what I did in Dafoe.

We were able to mix with the students on the station, but airwomen who weren't officers weren't allowed to mix with the male officers, but a lot of them did.

Airwoman: I was an Assistant Section Officer and I was given the Precision Squad to take on tour recruiting. There were 55 airwomen, girls from all over Canada and Newfoundland. There were cooks and GDs and everything else in the trades, and we toured the whole country. They did 287 movements without a word of command. After a year we had recruited 17,500 women, and we had filled our quota.

Some of the guys had cars or we could make a deal with one of the civilian instructors to rent their 1930 Essex or Star, one of those types; nobody had anything fancy. Gas was the hardest thing to get because you had to have ration coupons, so sometimes we'd take gas from the aircraft. But the RCMP used to watch for that, and if you got caught with aviation gas in a car there was a pretty heavy fine.

Hitchhiking was a great experience. You could go anywhere, and we did go everywhere. You'd go out the front gate of the station not having a clue where to go, and you'd get a car that was only going as far as Brampton, and you'd say, "Fine, we'll go to the dance." If the car that came along was going to Toronto, you'd say, "OK, we'll go to Toronto." Or maybe he'd stop just on the outskirts of Toronto and the next car you'd get in they'd say, "We're going to Oshawa." So, "Good, we'll go to Oshawa."

I had twelve dollars in my pocket and I had three weeks leave, and we headed for Chicago. We got there by hitchhiking, stopped off in Detroit for a day and stayed with the Salvation Army—they were giving wonderful service to servicemen. We got to Chicago and checked into the YMCA Hotel —fifty cents a night. Go to the USO—meals cheap as can be.

An American sailor came over and said, "What are those uniforms?" . . . "RCAF." . . . "Oh, RCAF! Where are you from?" . . . "Calgary." . . . "I've got a cousin in Calgary; his father's the coroner." . . . "Yeah! I went to school with him!" And that was always good for an evening. I never had trouble meeting people.

We hitchhiked quite often on weekends. A whole gang of us would go, boys and girls together. Once there were about four of us girls at the highway and along came a hearse and stopped and picked us up, and they had a whole bunch of fellows in there from Clinton, and we drove into London where there were a lot of retired farmers. During the war there was religion around, and respect, and the hearse pulled up on the main street and these old farmers, nice gentlemen, all stood there and took off their hats and bowed their heads. We opened the door and *whooosh*, out piled a bunch of service people like a gang of crazy nuts!

Hitchhiking in those days was nothing. You'd go anywhere the car was going. You'd leave the station Friday night and you had to be back by eight o'clock Sunday night, so you'd leave pretty early Sunday, because you were depending on traffic and there wasn't what you have today. You might get a ride for fifty miles and then get bumped and have to stand for half an hour. But *everybody* would pick you up.

The Australians wanted to see New York, so I went on a train with a bunch of them and spent a couple of days seeing the big lights. We were treated beautifully, and the Aussies were fascinated by the masses of people. We were standing outside a theatre marquee looking at the advertisements and the billboards, and the manager came out and asked, "Would you boys like to see the show?" But the Aussies said, "No, thanks." They weren't particularly interested; they were enjoying watching the people.

We went to the Stage Door Canteen, which the theatrical world had set up in New York for entertaining people in uniform. On the wall they had boards listing all the stage shows and all the top line movies and all the various eating places. They had stars or numbers after them, and each day they'd get so many passes for the shows or for meals. So you made a choice; if you wanted to go and see show number nine and you wanted to have dinner at so and so, you put in a request. And if you were lucky and they still had tickets, you'd get dinner and a show courtesy of the operators.

They had a park at Brantford which had an outdoor dance hall in summer. You'd meet a girl and you'd have a few dances with her and usually she'd say, "Come on up for supper on Sunday." If her mother was a lousy cook, you'd dance with somebody else the next week and try your luck again.

I had a 48, and a car stopped with two gentlemen and they said, "Where are you going?" I said, "Where are *you* going?" They said, "Toronto." I said, "I'm going to Toronto, too!" It was the Fourth of July and the American Legion was having a good will invasion of Toronto, and these were two dentists from Detroit. We got chatting and, "Where are you staying?" I said, "Nowhere." So I stayed in their room for the whole weekend. They had an old convertible and I was in the parade with them, and they bought me my meals. And Sunday afternoon they drove me right back to the station gate.

We were hitchhiking back from Chicago and had damn little money on us. We were going to be late and thought, "Gee, to cover ourselves we'd better send a wire that we're going to be a couple of hours late." We found that the cost of the telegram would be about what we had in our pockets, so we wired the CO, "Anticipate a few hours late . . . anxious to fly . . . looking to good meal." And he accepted that as a couple of fellows heading for school. As long as we were on parade the next morning, that's all he gave a care.

It was hard to get into Toronto. You had to rely on a hitchhike from the station because there was no traffic until Malton, and from Malton it was another hitchhike. The odd time you could go over to the Trans-Canada Air Lines hangar, and if the limousine had only three passengers they might take you. But it was a hard job getting into Toronto.

At Christmas 1940 they gave us a 48, and three of us came home to Edmonton. We stayed the whole 48, and of course, going back we were AWOL. We thought we were doing pretty good, but all of a sudden the train stopped in Winnipeg and there were a bunch of SP characters got on. "OK, you're on a 48 and you're AWOL, and you're not even back to Montreal yet." . . . "Well, we're just on our way back and we can't get back any faster." So he said, "Well, seeing you're on your way, go ahead." We got back there and they put us in the digger for seven days. They wouldn't give us sheets or anything; you just slept on the springs with a blanket. You couldn't read, and every now and again they'd have an inspection and they'd go through your bunk to see if you had any magazines or anything. And if you did, you were given some more time.

We used to make all our own fun and we had the damnedest bunch of talent on the station. We had a concert group called the Rip Cords that were in tremendous demand and used to travel to all the different stations. When they went to Winnipeg the biggest theatre in town would be absolutely jammed. It was crude humour, but it was well done.

I was playing the piano in the Officers' Mess one night, and we had a visit from Buzz Beurling. He was quite a guy at that time. I looked up and saw who it was, and he had a big smile on his face, and gee, it just lit me up. He said, "You'd make a great fighter pilot; your co-ordination is tremendous." But of course, that never happened!

I was playing the piano and the CFI was behind the bar. I used to drink rum and coke, and when I was finished the boys would go back and get me another. But this CFI started putting liqueur in along with the rum, and, oh, God, was I polluted! I can remember running out of the mess to put out a fire in a Lysander that was out in the rain. It took four guys to hold me back and put me to bed. I never woke up till 11 o'clock the next morning, and I thought, "Holy Jeeesus! The stink!" I had thrown up rum and coke and liqueur all over the floor, over the whole bed, over myself . . . and when I breathed!

In Dafoe you were really isolated from everything. The closest place was Boomtown, about fifteen miles away, and that was where the CN came in, so we had our recreation on the station. We used to put on concerts, and I would make costumes for the other girls because I worked in the parachute section. We had bowling there and we had badminton, and almost any kind of sports. We also had bikes supplied by the Air Force which we were able to take off the station, but we had to be sure they were back by ten o'clock.

I never drank, but there were a few girls that would drink, and when I turned 21 they would say, "How about coming into Boomtown and picking up a bottle of goof?"—which was wine; we called it goof. You were rationed; you were only allowed one bottle per person, so I would get a bottle and give it to them.

We had the weekend off. "Let's go back into the province and see how the Quebec folks live; let's go up to Grande Mère. Hell, there's no military establishment around there—no Air Force—no Navy. That should be a good place to go for the weekend."

We went into some kind of a pub where we could get a beer. It was a cold looking place, but we could sidle up to the bar. So we hung our coats and hats on hooks and sat having a drink.

Some other fellows came in, and we didn't pay much attention, but then we looked around and, by golly, our coats were on the ground and their coats were on the hooks. So we thought, "T'hell with this," and we took their coats and put them on the floor and we put ours back up! There was a bit of mumbling and a few more Frenchmen came in, and they started to crowd us. We said, "Come on outside. We'll take you fellows on, any one of you, one for one." We could see that we were outnumbered and we didn't know what these guys were gonna do. There was a picket fence, so we each ripped a picket off and backed ourselves against a brick wall and said, "OK, we'll get you comin'."

The police came along and said, "We don't know why you fellows come here all the time. You're always picking fights." . . . "Well, gosh, we wanted to come here to see what it was like in Quebec." Apparently this happened quite often, but they couldn't understand.

One night four or five of us went into town and ended up tanked with booze in the local cafe. One of the lads was a very well educated French-speaking fellow, and all of a sudden he got up and flailed into some civilians sitting in the booth opposite us. These French boys didn't know that any of us spoke French, and this one fellow who understood took exception to what they were saying about turkeys in the Air Force . . . and the fight was on.

We didn't worry about the fact that they couldn't speak English down in Victoriaville—we kind of accepted that—but we did find it shocking that three Canadian airmen standing on a street corner were told to move, because it was against the law to stand three in the street. Here we were, in the uniform of the Royal Canadian Air Force, and told by a policeman to move because we weren't allowed to congregate. Here *we* were joining up. They evidently weren't fighting for *their* bloody country, France was overrun. What the hell, this wasn't an English war, this was a *World* War, and their mother country was overrun. To me that was strange.

The civilian girls would come out to dances in the Drill Hall. Most of the Air Force stations had their own entertainment, and we had our own band. We had guys who were pianists, trumpeters, and one of them used to play just as good as Harry James; he could really make that trumpet talk. I was a drummer.

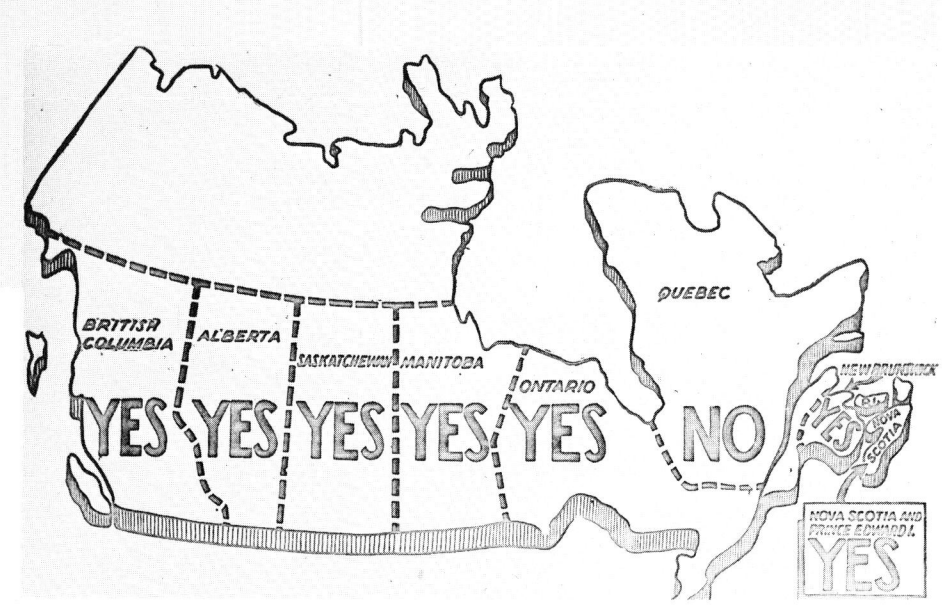

Victoriaville ITS was my introduction to Quebec. I was there about the time we were voting for conscription, and we used to go around with razor blades scraping the signs off the windows, "Vote Non". The Air Force was filling the training schools in Quebec with people from everywhere else in Canada, but we had only one Frenchman on our course. It was then that we realized there weren't too many French Canadians that were pulling their share—lots of factories making war goods, but damn few French fellows joining the flying service.

When we were in Quebec there was a great deal of resentment against conscription. The monseigneur of the local church got quite political about it at mass, which didn't help to cool tempers. The only French people that the Air Force got were the ones that made it in before they got drafted into the Army.

There were about a thousand men in the hangar in double decker bunks, and about 300 of them were Frenchmen trying to avoid conscription. It was coming up in Parliament, and they were afraid they were going to be drafted, so they had all joined the Air Force as groundcrew. The night they had the vote these guys were noisy as hell—they'd been drinking and they wouldn't shut up. We all waited until they'd quietened down and went to sleep, and then took a bunch of their double deckers out through the hangar doors and onto the field and left them there. And hoped that it would rain!

More than 15,000 aircraft of twenty different types were put into service with the RCAF between 1940 and 1945 specifically for use in the BCATP. Initially, a large number of these were supplied by the RAF, and some had seen hard service before their arrival in Canada. As the Plan reached its pinnacle, some stations were flying 14,000 hours a month, and maintenance and repair became important factors for smooth operations.

Fortunately, Canadians had served their apprenticeship in aircraft maintenance. For almost twenty years the bush operators had looked to the small shops to carry out regular running repairs and engine overhauls. These had been sufficient only to handle the bush requirements of the '20s and '30s, but they still provided the base for an industry.

When the British moved their schools to Canada they quickly learned that it was going to take more than RAF personnel to assemble and overhaul the machines they had brought with them, and the RCAF found themselves in a similar position. BCATP organizers turned to the commerical repair shops to carry out the assembly, modification, and

OPPOSITE:
At Christmas the big deal was comradeship with your fellow erks. Generally there was a parade to the Mess Hall and the food was extra good, and there was beer and so on around to drink. Some of the guys would have a little tree in their barracks, and if you had presents you'd put them around it; you'd do the same as you'd do at home. We'd invite friends in and have a drink or two. It was fun.

major maintenance work of most of their aircraft. The shops grew to handle the tasks required of them. The innovative engineering and dedication of the Canadian repair shop industry was to a very large degree responsible for the success of the BCATP.

Shop Foreman, industry: When the Commonwealth Air Training Plan got started they were looking for people to assemble some British built aircraft, and that's how Macdonald Brothers got into it.

The original Ansons came in large wooden packing cases, and I guess they were top-ship loading, because they were pretty weatherbeaten by the time they got to Winnipeg. I can remember a lot of problems even off-loading them from railway flatcars, because there wasn't any equipment around for that job. We had a group of people which we called the bull gang, and we rigged up a derrick arrangement and finally got them off almost by hand. We put together quite a few of those.

Shop Inspector, industry: There's an apartment up on 124th Street and 5th Avenue in Edmonton that was built out of Anson crates. One of the inspectors made arrangements to buy them and they built a horseshoe-shaped block. There must be twelve apartments in there.

Inspector, industry: Some of the wings of the old Ansons warped very badly. The moisture that was contained in those crates caused a good deal of the problem, and on some of the aeroplanes, to make them fly properly, we had to shim the flap down on one side. If they were flying wing low, there was no way of correcting it with trim tabs—it was often far beyond trim tab correction—so you had to shim the flap down, and that became a permanent part of the wing. And of course, that would slow the old devil up considerably.

Shop foreman, industry: We had Fairey Battles that came over the same way as the Ansons, and we had to assemble them. In those days the Battle was a modern aircraft, but it flew like a rock if the engine quit. All the thread sizes and the standards and everything else were UK and almost unknown in Canada. Spare parts were non-existent, so we had to sort of rob from one to the other to get them out the door. A lot of parts we made. They may not have been strictly to drawings, but we had some small machine shops and we had people that could do some welding and make sheet metal parts and things like that—aeroplanes weren't all that sophisticated in those days. Later on we got involved in an overhaul operation where the aircraft were completely disassembled to nuts and bolts and bits and pieces, and they all went through reconditioning and were put back together as almost new machines.

Shop Foreman, industry: In Edmonton we had not only the Fairey Battles, but the Ansons and the Airspeed Oxfords and later the Harvards. The Oxfords were flown in because they had been in operation and they were here to be reconditioned.

We had Ansons parked all over the outside of the station because we didn't have room for them in the hangars. They're all wooden aeroplanes, and they got condensation in the wings and the glue would grow mildew and all kinds of stuff. You could open a hatch and stick your nose inside, and if it smelled like cheese you'd know it had gone too far and you had to abandon the thing.

Inspector, industry: Those Oxfords were very bad for moisture. They used to trap water in the wings, and if you pulled a certain panel off you'd have to stand back, because about five gallons of water would come out. It seemed like the glue just fell away and plugged the holes, and then the moisture gathered. They were very bad for dry rot.

At that time I was only 20 or 21 years old and I guess I had a lot of brass, because if I thought a wing was no good I'd just throw it out. I was making some pretty big decisions on my own, but there wasn't anybody there that was any wiser, really—we were all learning together. At that time there were around 500 carpenters working on aeroplanes, and as time went on it just grew to three 8-hour shifts. Everybody learned. We had guys and gals off farms and everybody got along well because *nobody* knew *anything,* so everybody helped everybody else. We even had blind people there splicing cables just like a sailor does with a rope. It always amazed me how they could splice those cables, and wrap them with string, and lacquer them afterwards, without being able to see.

Ground Staff, industry: We went anywhere they needed help. They tried to maintain an 80% serviceability on these stations, but sometimes they had accidents and they had to contact us. At that time we were the biggest repair depot of that type in North America.

I was sent down to Lethbridge, and I was trying to put a Bolingbroke together by myself. I'd never worked on one. I didn't even have a parts book; all I had was my little tool box, and I was stuck in the back corner of a hangar. I had to put in a new fin and rudder, and one elevator was completely gone and the other one was badly damaged. I took everything off I could think of and I'd shake the fin post, and it wouldn't come; it was just like it was cemented in there, and I couldn't find anything holding it. This airman came along and said, "I'll bet you did like I did once. I forgot to take out those bolts that go through the . . ." and of course I had to make like, "Oh, yeah, I forgot about those." But really I didn't know they were even there!

Then I had to rig it. I had no information, so when nobody was in the hangar I'd drop a plumb bob from the rudder and measure along the floor, and then I'd go and do the same with half a dozen other Bolingbrokes in the hangar, and set mine on an average with them. The same with the elevator—there was nothing to measure from, so I'd look at different aircraft that had about the same amount of air in the oleos and in the tires, then I'd go over and lift my elevator up and measure to the floor, and the same when it was down, and I'd set the stops. To check the cable tensions I just measured with my fingers. I finally got it together and they shoved it outside.

They had a policy there that if you repaired an aircraft you were to go on the test flight. I'd never worked on a Bolingbroke and I'd put the thing together with my eyeballs, and every morning the damn thing's still sittin' there. I'd sweat blood all day, and every time I saw an airman coming I'd think, "Oh, Oh. Here we go." But one morning it was gone!

I often think about that—how crazy the company was to send somebody twenty years old, all by himself with just a tool box, to put on a whole new tail. You know, you wonder how we won the war.

Official, Dept. of Finance: The most serious equipment problem that we ran into was after Dunkirk in 1940. It was clear that the Brits were going to want all the aeroplanes that they could possibly produce to be fighters to defend Britain, and so they were not going to be sending any more aircraft to Canada. This meant that we had to go out and buy planes where we could get them rather than having them supplied by the British as part of their contribution to the Plan.

The Mark V was a completely different machine from the earlier Ansons, somehow roomier and much more pleasant all around, although a lot of the pilots used to say they liked the old aeroplanes better. But power-wise, there was certainly no comparison.

IT TOOK QUITE A WHILE TO BUILD AN AIRCRAFT

Management, industry: Federal Aircraft were sort of the central paperwork group. The contracts were let out and Federal looked after the engineering and all the design changes. We had people from Federal Aircraft in Winnipeg all time. They were there to expedite the supply of materials and things like that.

Official, Dept of Finance: Federal Aircraft and the Anson V program was not one of our happiest experiences. They moved quickly to get it going, but we were slow in getting the production up to the capacity that had at first been expected.

Prime Minister's Staff Member: Ray Lawson was the head of Federal Aircraft, and a lot of people questioned his capability, but he was a very good and successful businessman. The whole trouble was that everybody expected the aircraft to be turned out a week after the company was started. It took quite a little while to build an aircraft.

There were no replacements readily available for the Avro Ansons. They were trustworthy and versatile trainers that were being put to a multitude of different tasks, and although they had their problems with age and weather, they were nevertheless performing duties that could not be carried out by any other twin-engine aircraft in use at the time. Suitable engines for a newly designed replacement could be purchased in the United States, and it was thought that a variety of sections for a new airframe to carry the engines could be built by a number of Canadian manufacturers. The Department of Munitions and Supply was given the task of co-ordinating the project to build what became known as the Anson Mk V.

Official, Munitions and Supply: We had a division of aircraft production in the Department of Munitions and Supply which was headed by R.P. Bell, and it was decided to set up a separate company which would be responsible for the Anson V. It was called Federal Aircraft and it was headed by Ray Lawson from London, Ontario, and he was supported by a man by the name of Newman, who was from the engineering department of the Canadian Pacific Railway.

Ground Staff, industry: When the Mk Vs got into the act they departed from the old tubular steel fuselage with fabric covering. It became the same type of wooden structure as the wing, with fabric covering over the wood to protect it. And the aeroplane was now equipped with Wasp Junior engines and a hydraulic system that would retract the undercarriage.

AOS Flight Supervisor, civilian: There used to be a lot of Americans coming through Edmonton on their way to Alaska with all types of aircraft. We'd look after them and they'd headquarter at the Air Observer School. This Texan came wandering over one day and looked at one of the Ansons, and he put his thumbnail on the skin of the wing, shook his head, and said, "Just as I suspected—caaade-booad!"

THE CRUNCHES & THE CRASHES

There was a fellow on our course who wasn't the world's best pilot, but he tried valiantly. Unfortunately, he stalled short of the field and came down into a couple of elms, and just sat there like a kid's kite twenty-odd feet off the ground. He suffered the ignominy of having to be rescued by the local fire department and their ladder truck.

FACING THE FACTS

Instructor, RAF: One night I was taxiing with a pupil in a Cornell, getting him to taxi properly so that he could see where he was going. All of a sudden there was the most awful impact behind me—bang!—and the aircraft was driven forward. Another aircraft with a student and an instructor in a hurry to get circuits done had been taxiing at full speed and had run right into the back of us. It had mounted our aircraft and the prop had just stuck in the wood behind my seat! Slashed the tail right off. That was the nearest I came to getting my head cut off. It was awful! The prop had just churned its way through that plywood.

Ground Staff, civilian: There was one thing about the old Tiger Moths—when they went on fire, they *really* went on fire. Because of the way the gas tank was held on the struts, if there was a heavy impact, those struts went right through the tank. And then she blew.

Staff Pilot, RCAF: We used to have exercises to train the navigators how to map read from low level. Those flights were supposed to be between 300 and 500 square feet, but we were often right in the tree tops. When you got down to 300, you just couldn't resist getting *right* down, and that sometimes led to problems.

Student Pilot, RCAF: I was about ten miles from the airport when I lost an engine, and I never quite got back to the damned station. I landed on the prairie and it turned upside down and I broke my back. They put me in a plaster cast for six months.

Usually it was just a motor failure or a student got lost and ran out of gas and had to land on rough terrain or deep snow or bushes. A lot of it was inexperienced pilots; maybe a student would come in and hit, and then bounce, and stall, or he would ground loop.

Ground Staff, civilian:
We'd have two or three little crashes a day; sometimes out of twenty aircraft we'd have five going. Some of the guys didn't give enough clearance when they were taxiing around and they'd smash wing tips, and the odd time when they were running up on the ice the chocks would let go and the student wouldn't realize he was moving . . . and wham! That was the end of two aircraft.

Student Pilot, RCAF: *When we started our flying training it was in the late fall, and the ground was frozen just like cement. But then we got a bit of a wet snowfall and one of the Moths went on its back when it landed. They hadn't got any of them on skis yet, and I guess they figured they'd just carry on with the wheels, but that ended that day's operations right there. Nobody got hurt, but it's sure hard on aeroplanes.*

Officer, RNAF: *The P-36 was really a fighter aeroplane, and the engine seemed to be bigger than the aircraft. We did advanced training on those off the Island Airport.*

Ground Staff, RCAF: *If they were forced down, then salvaging them was a big problem. At the start we didn't have too much equipment; in Saskatoon we had a truck and a long piece of rope. The first one we had to go after was a Harvard that went down in a stubble field; it had made a belly landing. We had nothing to lift the aircraft with, so we had to dig a hole underneath big enough to pump the undercarriage down. There was no room for shovels, so all we could do was use screwdrivers and hammers to chip the mud and clay out, and that's hard land out there. We managed finally to get the gear down, and then we put the truck in there with the rope, and we were able to pull it out of the hole.*

Student Pilot, RCAF: *Those old Ansons must have been built better than they looked. We had a guy land his on top of another on the approach, and somehow the fellow in the lower one managed to get the two of them down OK. When I was on guard duty before ITS, we had to look after a bunch of them that were stored in pieces under the grandstand at the CNE in Toronto. But there must have been something wrong with them, because after guarding them all winter we had to move them all out into the field and they burned them.*

Junior Officer, RAF: *We had some American student pilots that came up to work as staff pilots. They gave a heck of a line about how much experience they'd had and all this kind of stuff, but they were hopeless. They were very willing, but their flying ability was non-existent. One of these guys saw a storm coming and he figured he couldn't get back to the station, so he just landed in a clearing.*

AOS Staff Pilot, civilian: *One night they sent us north out of Quebec City—there's no lights, just bush. It was all instrument flying, and half an hour out we ran into a thunderstorm. On that night, out of twenty aircraft, only five of us found the station. Two went into the St. Lawrence River, others landed in the States, and some went down in the bush. So they didn't send us north too often.*

OPPOSITE:
Student Pilot, RCAF: *One Tiger Moth accident that we had was an American lad who had joined the RCAF. We were waiting in the flight room for our instructors and assignment of aircraft, and he spun in just off the edge of the field. The Medical Officer went whistling over, but when he got there the student was sitting on a stump. All he had were a few cuts and bruises.*

Student Pilot, RCAF: *We had a violent accident. The instructor and student went up to do some instrument work on a particularly rough day. Somehow things got out of hand, and they augered in from a spiral dive. There was just a bag of linen with a bunch of stuff in it. My instructor flew me over to see what was left, and the message was, "Watch it in a spiral; things can get out of hand pretty quickly."*

Student WAG, RCAF: *There had been a mid-air collision, and two planes had come down on a farm just outside Ottawa. We were detailed to go and guard the wreckage and keep souvenir hunters away until the investigating crew could go in and figure out the cause of the crash. One was a Harvard that had gone straight into the ground and folded like an accordian. The other was an Anson that was carrying an instructor and four students, and two of them had baled out, but they were too low, so they hit the ground and were killed. The whole thing had just disintegrated; it was an awful mess. And here we were, brand new recruits—we'd only been in the Air Force for a month—and we were supposed to guard this thing. They put us up in farmhouses, and eventually they cleared the wreckage. They poured gasoline over most of it and burned it, but it sure shook us up. I'll never forget the smell of the oil, and the gasoline, and the bodies. Every time I get into an aeroplane, I can still smell it.*

Ground Staff, RAF: *Depending on the rank of the person that died, there were so many men required for the funeral, so many pallbearers, and the Padre. If it was wintertime, the parade would be held in the hangar, and then the body transported to the graveyard. There was usually a Union Jack, and the officer's cap or the student's cap, then prior to the casket being lowered, the Union Jack was taken off, and so was the hat.*

7

An impression of sheer horror
Getting away with it
It could be pretty damned ripe!
Howling like ding bats
A very solemn affair
I couldn't see inside the cockpit
Nine months of retraining
Lamb chops with frills
I thought we'd turned over
Loose teeth and picket fences
The rest of the week I dug graves
The clock that was five minutes slow
They were giving us a choice!
Three shots for our side
Toting up the bill

THE REWARDS & THE REALITIES

There were almost two dozen aircraft types of varying sizes and configurations filling the flight lines of the BCATP schools, and whether a person was a pilot or a bombaimer, a navigator or a gunner, a maintenance engineer or a civilian, each type played its own nostalgic role.

For the most part these aeroplanes have their places in books on aviation history, and scattered examples have been carefully restored to join the ranks of honour in museums around the world. But there was one aircraft type that even today draws gatherings of admirers when it turns up at fly-in breakfasts and country air shows, and whether a student pilot was slated for fighters or bombers, the memories of that aircraft and its role in the Plan will be with him forever.

Civilians, too, remember it, for if they were not in the Air Force its snarl annoyed them, but if they were nearing the magic age of eighteen, those sounds would turn their eyes to the skies and lift their imaginations to the cockpits of Hurricanes and Spitfires, and heroic dogfights amid the towering cumulus. There is little doubt that of all the aircraft that were used in the Commonwealth Air Training Plan, none has had a more lasting or widespread nostalgic effect than the sight or the sound of a North American Harvard.

I remember walking through the hangar and seeing these great, bit, yellow Harvards. They looked massive. They had a 42-foot wing span or something —great, big thundering monsters. And I wondered, "God! Will I ever fly one of these?"

My first impression of the Harvard was one of sheer horror! I was scared to death of it. "*How* am I supposed to learn all this stuff?" And then, back to the books, looking at pictures of Harvard cockpits and trying to learn what all these things are.

I can remember, "H–T–M–P–C–G–S; that's the takeoff check. H is heat, T is throttle, M is mixture control, P is pitch full fine, C is carb heat, G for gas on, S for all switches on. That's forty years ago!

I remember the exhilaration of the first takeoff— the instructor sitting behind and me up front. And when he sort of gunned that big motor . . it was the first time I'd ever been up in an aeroplane that powerful. And that sort of push in the middle of your back . . . that sense of power that you had.

I was warned very early by my instructor, "For heaven's sake, don't let this thing get away with anything on landing."

You really had the power in front of you now, and the manoeuvrability was amazing. I was not allowed to move the controls unless I followed up by trim, and as soon as you trim that Harvard you can really fly it hands off.

I got to fly the Harvard when there's all the cumulus clouds, those fair-weather clouds where you could go up and cloud chase, which is just a *ball* with a Harvard. We had a great time. I can remember the Harvard with great affection; it was a wonderful aeroplane to play in.

I did my low flying in the clouds. I enjoyed doing weather checks in the morning when you climbed up through the overcast and came out on top. When you were there by yourself, that was really something.

They weren't as smooth and fluttery as a Fleet; a Harvard came 'round pretty fast. A Fleet was like a ballet dancer, where a Harvard, once you got over the top, was more like a stone.

I'd never had flaps before, and they gave it stability. I could just *feel* it, just *sit* on that approach. You had great control over it, and for the first time I was allowed to use power on the approach.

I remember doing precautionary landings where you go over the fence, and I used to think, "God, how can these things fly so *slow?*" We'd be doing 55 or 60 with the wheels down and full flaps, and you could fly along the ground like that, just over the fences.

The Mk II Harvard had square tips, but the Mk I was a little harder to fly. If you were really trying to make a three-point landing, the two wings wouldn't stall at just the same time, usually one would drop, and a lot of fellows caught those wings and hurt themselves. But that was a characteristic of the aeroplane.

They would drop a wing very quickly, and you really had to fly the thing till you were stopped on the runway.

I learned from the Norwegian instructors, "Don't have your tail too high, but have it off the ground and make sure that you plant your main wheels *on* the ground, and then throttle back very slowly. Let your tail come down and make sure that you have immediate rudder control over the tail wheel."

We always set the DI—directional indicator—to zero on final, no matter what the heading was, and as soon as we had any problem when the tail went down we went on the DI and just held the zero. That was something that we learned very early.

One thing that was emphasized over and over again was, "Never feel that you have completed your landing until you have actually turned off the runway." Because even though you're rolling down the runway with the tail wheel in contact with the ground, if you don't watch it very carefully, you could get into a ground loop so quickly that you just couldn't recover.

I was sitting in my aeroplane in the line-up for takeoff, and this Harvard came in and touched on one wheel. He had a bit of drift on and he was soon going across the runway and was very close to hitting the wing on the ground. He managed to get it back up, but he over controlled and he was then in a steep turn going in the opposite direction, and still only a few feet off the ground and right in front of me. But finally he got control.

I learned a great lesson at Trenton one night when I was flying by myself. I took off and when I was climbing out over the Bay of Quinte nothing was happening. I wasn't climbing very fast, and it was dark, and the engine didn't sound right. I thought, "Gees, don't hold the nose up." I was losing airspeed. "You're not gonna climb if you haven't got any power." It wasn't quite at the stall—just a mushing position—when I realized I hadn't tightened up my throttle quadrant, and everything had slipped back. I probably broke a few windows, because that Harvard went into full fine pitch and full throttle. But I never took off again without tightening that quadrant.

You could do anything in a Harvard; it was unbreakable. Noisy, but terribly strong. You could throw that thing all over. You never heard of anything happening to a Harvard through structural damage.

You could bounce it all over the aerodrome without it falling apart. It had a good feel to it; I was really impressed. This is quite a chunk of machinery, not just a flutter in the breeze like a Tiger Moth.

We could do things that we had never done before, like a quick roll. You would just jam your controls over to one side and *flip*—you're all the way around.

I remember climbing on board the Harvard and sitting inside and looking at the instrument panel—and that big radial in front of you—"My God, there's no way I'll ever fly this thing!" It was just too big; too complicated.

The Harvard had a lot of knobs and levers and instruments and gadgets in it that we weren't accustomed to. The little Fleet Finch had about four or five instruments and a throttle, so when you looked at this array it kind of made you think, "Can I handle this?"

We were doing neat stuff. They would send you out for an hour's aerobatics and you didn't have to pay a cent and you didn't have to worry about gas! It was a lovely aeroplane.

They were great things to aerobat from the back. When you're doing aerobatics it's like firing a gun; you need a gun-sight to get on the target. When you're in the front you don't have anything but this great big engine. But from the back, you'd just look along the bow of the aircraft, and it was so easy to roll them.

You can put the nose up just above the horizon and you can do three complete rolls, just like that. Just hold your nose right on that spot.

Without touching the throttle we'd take her tighter and tighter and tighter. While we were losing speed she'd start falling and you'd wait for a flick, and if she didn't flick you'd kick the rudder, and she will stall out on a flick roll.

If you weren't well strapped in you'd hit your head an awful bang; it just rattled around in there. It would flick on top of a loop and you were into a spin before you knew where you were. But those were the bad characteristics that made you alert, like a girl on a breezy day at an intersection has to be ready to hold her skirt down as she goes round the corner. It was like that in a Harvard; you had to anticipate that it was going to misbehave.

We had a little wee short guy, and he said to me one day, "Have you ever done a vertical reverse?" I said, "What the hell's that?" He said, "You fly straight and level and build up your speed, and then just do a tight turn. Once you get it real tight, a really steep turn, then pull back on the stick, all the way, and you're goin' backwards!" He said, "It'll turn right around and you're *going* from the direction you're *coming* from—you're going backwards." I said, "Isn't that kinda hard on the aeroplane?" He said, "Yeah, but, oh, it's fun!"

So being stupid, I tried it. Gees, what a rude awakening! All the bloody dirt came outta the floor, and oh, gees, the engine coughed, but I *was* going the other way, and I thought, "Gees, it works! But, oh God, it's got to be hard on the aeroplane."

I knew of people who did bunts—an outside loop—in a Harvard, but I never had the courage to try that.

Word got around that a Harvard could not be spun inverted, that you couldn't invert the aeroplane and then get it into a spin. But my instructor inverted one and actually forced it into an inverted spin, and he spun down five or six thousand feet before he recovered, in an inverted condition. When he told the other instructors that he had done this they wouldn't believe him, so he took up one instructor after another, a half a dozen of them in a period of a week, and each time put this particular aeroplane into an inverted spin. Of course, they were convinced.

The wisdom was that it was impossible to spin the Harvard inverted. But all of a sudden we got a memorandum from Training Command Headquarters: "We were wrong. The Harvard will spin inverted." Everybody *knew* that; there were all kinds of Harvards found in the prairies, inverted. But after we got the word I tried it a few times, and it's most uncomfortable.

All of a sudden you'd turn it upside down and straight down, and get to about 250 or 260 on the clock, and then pull it straight back and do a loop or a roll off the top. You'd wait to see the ground coming around, and then you'd roll it. Or you'd do a straight loop, or a vertical roll. That's what I used to like doing—just taking the thing upside down and letting it go, and then pull it straight up. And you could roll it about one and a half times before the speed dropped off and it fell out of the sky.

There was only one way to make it spin inverted, and that was off a roll off the top. You checked about where you would start rolling off anyway—deliberately checked yourself. Then stick hard forward, and slam on full rudder and hold everything right there. She went through some weird gyrations. One wing would fall off and you would seem to sort of do a complete *whrrooowmm*, until you were upside down and the nose would drop, and you were spinning, inverted.

There's a certain amount of gravitational pull so you're falling into your straps and you're not too comfortable. You don't feel you have the same control of the aeroplane. Your feet seem to want to come off the rudders, and when the spinning starts you somehow can't sense where the devil you are. You feel that you're alive, but you don't know where you're at!

It was awfully hard to recover without half looping out. The theory was, "stick right back and full opposite rudder." Then when the spinning stops, centre your rudder and push the stick forward, and you recover inverted. Try it! There's that recovering inverted bit that's so difficult. You have to have enough altitude to be able to half loop out, and by the time the spinning stops, you're going downwards, inverted, at a very high rate of knots, and you red out. It's much like blacking out, except that everything goes red, and your vision goes. It was a very, very dangerous manoeuvre.

It's hard to describe this tendency to get a red out. It's a peculiar feeling. You kinda see red through your eyes rather than having that black out feeling of losing consciousness.

If you got into it accidentally, by the time you realized what was going on and thought out the recovery of half looping, you'd better have about 10,000 feet available. I was an instructor at the time we got the word that the Harvard would spin inverted, and it was more or less obligatory to go and find out. They never did say anything about how they found out that it could happen, but it's not a pleasant sensation.

The Harvard wasn't an easy aeroplane to fly. It was a very sturdy aeroplane, and a good aeroplane for single engine instruction, but noisy as goddammed hell—no question about it. There's no other way of describing it!

At two o'clock in the morning coming back from night flying you'd give the throttle a couple of surges and the Harvard would give its traditional "Wooow, woooww"—which was the usual signal that you had promised your girlfriend you would give when you were coming in.

We had a couple of funerals for guys that got disoriented. In a Harvard there was no hope at all—it was game over. You just didn't crack up a Harvard and get away with it.

In pilot training, "getting away with it" was the game. There were unavoidable accidents, some through errors in judgment, others through mechanical failures, and still others because of weather. But by far the greatest number of flying accidents were caused by the games that were played by students and instructors alike.

Instructor, RAF: If they caught you low flying they just gave you a ticket out of the country and there were no questions asked. Unauthorized low flying . . . Bang!—and you were gone. When you became an instructor here was the chance. You had *carte blanche* with what you could do with pupils, and there was such a thing as instructing them in low flying—it was part of their training.

Student Pilot, RCAF: We used to beat up the beach. We didn't get too close to shore, because there was always some high official around. So we'd take the Harvards just far enough out that they couldn't read the numbers on them. We'd take them right down and across the water, and then you'd pull up, and away you'd go.

Station CO, RCAF: They did quite a lot of that. We were pretty strict on it, but people did low fly; you couldn't stop them. I was down at my cottage on Lake Winnipeg, and somebody came over low flying over the beach—probably over a girlfriend's—and I got the number, phoned the station, and got his name.

But we didn't treat those people *too* seriously. In the summer we'd make them put on their full flying suits and a helmet and a parachute, and they'd have to march up and down in front of the mess for an hour or two every day for a week, while everybody was going in and out for meals. They didn't like that very much, but that was about the most we did.

Student Pilot, RCAF: I knew the country well; all the landmarks were very clear to me. It was a beautiful day and I thought, "This is the time for me to go back and pay my respects to my old school." It would be fun to make a little low pass

Instructor, RCAF: *We used to play follow the leader in Cessna Cranes. It was in the fall when the larch loses its needles, and they were sticking up above the tops of the pine, but with no needles our leader didn't see them. All four aircraft got damaged. The lead's belly was completely torn out, and some of them had wing damage, but fortunately they all got back. His punishment was a posting to Kingston to instruct British Sea Cadets. To be sent down there was considered the lowest step in the Air Force.*

and get my friends to come out and wave. I thought it would be a rather touching moment. So from about 3,000 feet I lined up on the school, eased off the throttle, and put the little Fleet into a gentle dive. When I got down as low as I dared—which was pretty low—I opened the throttle and shot between the main school building and the covered rink. I remember passing between them very low, and as I revved the motor to climb out of there I saw people running into the quadrangle where we used to do our Cadet Corps drills. I circled around and lined up on the rink roof and made a second pass, and maybe got a little bit lower. Some stories have it that I actually touched my wheels on the roof, but unless I did it very smoothly, I wasn't aware of it. I didn't *intend* to do it. Fortunately, nobody reported me. I certainly wasn't pulled on the mat as a result of it. If word did get around, it didn't get back to the station.

Eyewitness, school boy: I was a very impressionable young teenager. We used to have to play soccer every afternoon, but I was much more interested in all the yellow aeroplanes that used to fool around overhead than I was in kicking a soccer ball. It was the noise they made that attracted me.

We had just finished lunch and there was a tremendous roar. I knew right away what it was, but I had never heard the noise so close. I ran outside, and there was this aeroplane, just up over the piece of bush where we used to play. It was right down low and it was turning around, and I watched. He came back towards the school and right down low onto the rink roof. I could even see the pilot and I'd never seen a pilot before. But then there was a lot more noise from the engine and the plane was gone. That was really exciting!

Student Pilot, RCAF: My instructor's one aim in life was to scare everybody shitless, so he'd get rid of the Ansons and get sent overseas. He shot up a canoe on a lake, and the canoe capsized, and he got in dutch for that. Outside of Hamilton he bounced his wheels on the Queen Elizabeth Way behind a car and then landed in front of it and took off again, and got caught at that. One night that I was with him he said, "Let's leave the circuit." So we turned out our nav lights and bogged off. It was clear moonlight. We were up near Orangeville and he spotted a train and he said, "I have control!" So he whipped her over and flew away from the train about five miles, then turned and headed back along the tracks with one landing light on—right down even with the wires on the telegraph poles, goin' hell bent for this train. When we passed over it there was steam comin' out everywhere. I guess they'd thrown on every emergency brake they had.

Instructor, RAF: We'd sometimes do excessively low, low flying. We had a designated low flying area, and low flying was supposed to be 200 feet. Well, *that's* not low flying; low flying is *two* feet! The wheat probably gets blown backwards a bit, but I never heard anybody complain about what happens to the wheat. But don't do it near their cows!

Student Pilot, RCAF: There was one time I was with my instructor and we took out about half a mile of wheat—levelled it right off with the prop. Cut the wheat right down!

Staff Pilot, civilian: Several of the Ansons came back with wire wrapped all around their undergear; in fact, one of them had wire wrapped around the propellers, but managed to make it back to the field.

193

Instructor, civilian: One day we went out to the local resort beach; a couple of the instructors had girlfriends out there and we were going to put on a little performance for them, but we got kinda carried away. There were a couple of old ladies out on the end of the deck, and we put them right into the water. We tried to stay low so they wouldn't get our numbers, but word wasn't long getting back. We each came up with a $150 fine. We were working as civilians, but we were RCAF on leave without pay, and in those days, that's how the Air Force disciplined us EFTS instructors.

Station CO, RCAF: Mostly the kids didn't fool around, the instructors did. We had one fellow in Souris that came back with hay all through the cowlings and the engines. He just went so low that it swished up with the props. He was charged with low flying, but nobody really wanted to pin anybody down for that. Those guys were misunderstood and they should have been allowed to do it once in a while. It didn't really hurt anybody. It was usually their own neck, and most of them were frustrated anyway.

Student Pilot, RCAF: My instructor took me under the Ivy Lea Bridge. That was a thrill for a young pilot in training who wasn't supposed to learn to do things like that.

Instructor, RCAF: There was a railway nearby, and every morning there was a train at a certain time. So we had great fun getting down low, just at the fence level, and flying along and catching up with the train, and everybody'd look up at you as you'd fly by.

Instructor, civilian: Our real prime target would be a fellow on a haystack. There'd be a team of horses over on the other side, and we caused more runaways than you could ever imagine. Those poor guys—he'd be up on top waving his pitchfork and the horses would be long gone across the field. Then we'd catch a fellow out on a speeder or a hand car on the railway, and you could sure take them by surprise. You'd come up behind them and they'd bail right off, and let the old hand car go on by itself. They wouldn't know what had hit them, it was so sudden. No sound, and then we'd scare the hell right outta them!

Instructor, RAF: One of the exercises pupils were supposed to do was cross-country instrument flying. I would take them a long way out of the designated flying area and I'd give them a bunch of courses that would take them way up into the mountains, and I could enjoy the scenery. The mountain tops around there are ten or eleven thousand. Then with the pupil still under the hood

I'd put him on various courses and have him gently descend and descend, till he's at about 5,000 feet and in a valley, and then pop the hood on him.

Instructor, RCAF: I helped a farmer chase his horses into the barn. In Quebec they have these great, long, narrow farms, and one morning we spotted this farmer trying to catch his two horses, but he couldn't get them. They were running this way and the farm was that way. So we came in over the field and turned the horses around and chased them right back into the barn—I thought they were going to come out the other side. I don't know if the farmer really appreciated it!

Student Pilot, RCAF: The Saskatchewan River ran right past the airport at Prince Albert, and to prove their ability to their buddies—and maybe to us, too—the instructors would fly down the river and get their wheels wet, and then come in and land and show they'd had their wheels in the water.

Ground Staff, civilian: Harry Higgins's real ambition was to create a misdemeanour grave enough that he would be posted to active service, because this flying instructing thing was not his bag. He just hated it—he wanted to get into action.

It was the 24th of May, 1941. Higgins took an aircraft and as soon as he got airborne he made a pass at the TCA tower, then he did a few loops over the field, and then went over to Lethbridge. There was a golf course there, and on this holiday there were a lot of people, among them the manager of the Bank of Montreal and the manager of Eaton's. Higgins came right in on them, just as close as possible, and the Montreal man went right to the ground, but the Eaton's man, he jumped and went right in the water.

There was a great hullaballoo about this, and they had a court martial, and during the questioning Higgins's attorney asked, "Mr. Higgins, it wasn't necessary for the manager of Eaton's to jump into the water, because you would have missed him, wouldn't you?" Higgins thought about it for a moment. "No," he said, "I don't think I would have!"

Student Pilot, RAF: *The instructors we had were all ex-operational pilots who had finished a tour and been sent to Training Command to have a rest. Most of them loved close formation flying, and it wasn't unusual for a bunch of pals to get together and take a bunch of their students up with them while they played around—touching wing tips and this kind of stuff. And we would sit there thinking, "How did these guys ever learn to do* that?*"*

Student Pilot, RCAF: One instructor spent more time in the control tower than he did instructing. They'd put him in the tower for three days and wouldn't let him fly at all. The odd time those guys in the tower had to shoot a flare. The normal thing was to point it skyward and pull the trigger, and the flare would arc up and then back down again somewhere across the field. But this guy, he'd aim it right at the aircraft and bounce it in front of the pilot, and have a good laugh.

Student Pilot, RCAF: I'd just finished my twenty-hour check on Tigers, and at twenty hours you're feeling pretty good. There was a 3,000-foot cloud base, and the met people said that it would be topped at 5,000. We'd been under the hood, but I hadn't flown any real instruments, but my buddy and I were taking off at about the same time, so we said, "Why don't we meet up at a little over five and have a bit of a chase around."

I flew north out of Goderich and entered the cloud and started to pick up ice, but I was at about 4,500 and I thought, "I've only got 500 feet to go. I'd better get up." I got to around seven, and that Tiger Moth was just a snowball—it was just loaded with ice. I was on the verge of panic, but I had to get on top and hope to get rid of the ice, and then come on back down through, fast. Well, I got to 8,000 and I was still in the stuff, and then the aeroplane decided, "Enough of this nonsense!" It stalled, and I went into a spin. I knew I was in a spin, but I didn't know whether to bail out. To start with, I thought, "How am I gonna get outta this thing?" But the thing that I was really worried about was, "Gee, they'll wash me out if I bail out." So all theory went out the window right then and I got down to real basics. I did what you're supposed to do. I figured out which way I was spinning, and opposite rudder, and brought her down through the cloud. And when I landed I *still* had a real load of ice on that Tiger Moth.

You always had to go and report in, and I remember my instructor asking, "How'd it go?" And I said, "Fine." . . . "What did you do?" . . . "Oh, I did some steep turns and . . ." I didn't go up in cloud again for a long while!

Student Pilot, RCAF: We used to chase up and down the Saskatchewan River. We'd get down in there and practise shooting each other down with cameras, because we didn't have any drogues to shoot at. You'd watch to see if you could really get the guy down or how close you were before you pushed the button. It was great fun down inside the gully; it was about 100 feet deep, and we'd chase around in there.

Student Pilot, RCAF: My instructor said, "Go and get that Anson and taxi it around to the hangar; it's been damaged." It had a great smash in one leading edge, the props seemed to be bent, and the windscreen was all smashed.

An Australian instructor had been out with a student and they had been beating up along the Bow River. It's fun to go down the river, but there are cables across, and you sure gotta watch for them. They had hit a cable, and it skidded up the nose, smashed the windscreen on the instructor's side, went past the props and mangled them a bit, and then went through the leading edge into the main spar and broke off. The instructor's windscreen was so completely shattered that he thought it was game over, and in reacting he threw up his arms and let the controls go. The student managed to take over and climbed back up out of the Bow and went back. He said both engines were shaking, and it was vibrating like crazy all over.

Student Pilot, RCAF: I wanted to get through. I wanted to get my Wings, so I didn't really fool around very much, but a week before the presentation a chap took three feet off the end of a Harvard wing. Somehow he got the aircraft under control and got it back to the field, and that took a good deal of skill. But he was marched before the CO, set back a whole course, and confined to barracks for thirty days, because the accident had been the result of unauthorized low flying. But he was told, "If you'd been overseas, you'd have been praised for airmanship, and you'd have been awarded a DFC!"

Student Pilot, RCAF: I'd finished my sixty-hour check and my preliminary Wings check on Ansons, and the next exercise was a cross-country navigation. I was the navigator and Logan, who was an RAF chap, was the pilot.

We took off and got up around Owen Sound, and he said, "We might as well go down and see what's going on." We went scootin' along pretty low. I was trying to navigate, but enjoying the low flying just as much as he was . . . and we hit a tree! It hit with an awful wallop—just a terrible *wham!*—which brought us right back to the business at hand in a hurry. The aeroplane was *flying* fine, so we finished the exercise, but decided we'd better stop in the outfield and have a look, because if it was noticeable we'd have to make up some kind of a story. I climbed out and we had half a tree on the bottom of that aeroplane, so I pulled it off and threw it in the infield, climbed up to Logan, and said, "Obviously, we've done some damage. You're the captain; I guess you'd better report it."

Well, he didn't report it. At ten o'clock that night an armed guard arrived—"Come with us!"—and I'm marched to the Flight Commander's office. It was just like a trial. I hadn't seen Logan and I didn't know if he had already been there or if he'd told them a story; I had no idea. They marched me in and the Flight Commander was sitting there, and my instructor there, the Base Commander over here. And then they asked me if I'd been flying Anson 3305 in the afternoon. I said, "Yes, Sir." . . . "And what were you doing?" . . . "I was the navigator." . . . "Who was with you?" . . . "I was with LAC Logan." . . . "What was *he* doing?" . . . "He was the pilot." . . . "Your exercise was a normal navigational exercise?" . . . "Yes, Sir." . . . "Did you hit anything?" And I said, "Well, that's possible. I did hear a bang." . . . "Where were you?" I said, "There's an escarpment at Owen Sound. There was a lot of cloud around and to get a better pin point we had to go down a little lower, and LAC Logan was looking at something on the engines when the escarpment went up, and he realized it a bit too late, and we hit the top of a tree."

Logan got washed out immediately and he got sent back to England, and he was within two weeks of graduating. And they told me that if I hadn't finished that Wings test I would have been washed out, too. As it was, they said, "You're not going to get a commission." I thought, "I don't care about a commission. I want my Wings, that's all." But that's how close I came to not getting them.

Instructor, RCAF: The Harvard had a piss tube, a rubber tube down to an outlet in the fuselage. There were all sorts of convolutions in this tube, and there was always some trapped in there. When the student had the call the instructor would trim the aircraft very nose down, and just as he was taking back control he'd let go of the stick. The student was caught by surprise, and the Harvard would *thonk*, and the maps and everything would fly up into the top of the coupe top . . . along with all the piss in the tube. Depending on how long it had been there, it could be pretty damned ripe.

Student Pilot, RCAF: The first time I saw the pee tube I picked it up and started to shout into it! And they said, "No! You don't speak into *that!*" and everybody laughed. Nobody ever told you about things like that, you just found out. Your instructor didn't say, "If you want to have a leak, that's where you leak." They took a lot for granted.

WE HAD A RULE OF THUMB

Student Pilot, RCAF: It was a rather nice day; a few clouds in the west, but it didn't look bad. Just for the fun of it we were going to get a couple of Ansons and see how high we could get, so I drew one that had variable pitch props and George drew one with fixed, and off we went. I didn't have too much trouble getting up to 10,500, 11,000, something like that, and then the cloud was thickening up so I beat it off home and landed.

By God, it didn't take too long till it was really blizzardy, but we had a rule of thumb if you got lost: "Head west till you hit the lake, along the shore to Grand Bend, fly east to the railroad, and up or down the railroad you'll find the station." But George hadn't seen the squall coming till it was too late, and then he thought, "Better head for home." He started home, but after a bit—"Where the hell am I?" He was really getting low on gas; he couldn't tell where he was; he got lower and lower; the squall got thicker and thicker; and he finally landed in a field. He phoned the station and they said, "Where the hell are you?" He said, "I'm between the "O" and the "L" in Petrolia, and it looks like I'm going to be autographing parts of propellers all afternoon."

Staff Pilot, civilian: Over the turning points on our nav exercises a lot of us did stall turns. Instead of making a nice gradual turn onto the new heading, we'd pull the nose up in the air and kick rudder, and all the navigational instruments and pencils and rulers that they were using would be all over the cockpit.

Student Pilot, RCAF: In Ontario they used to have a lot of summer storms—huge cumulonimbus would build up—and it was great sport to have somebody trying to fly under the hood and keep giving him alterations in course until he flew into one of these. We used to think it was great sport to see this poor guy being tossed all over, and then

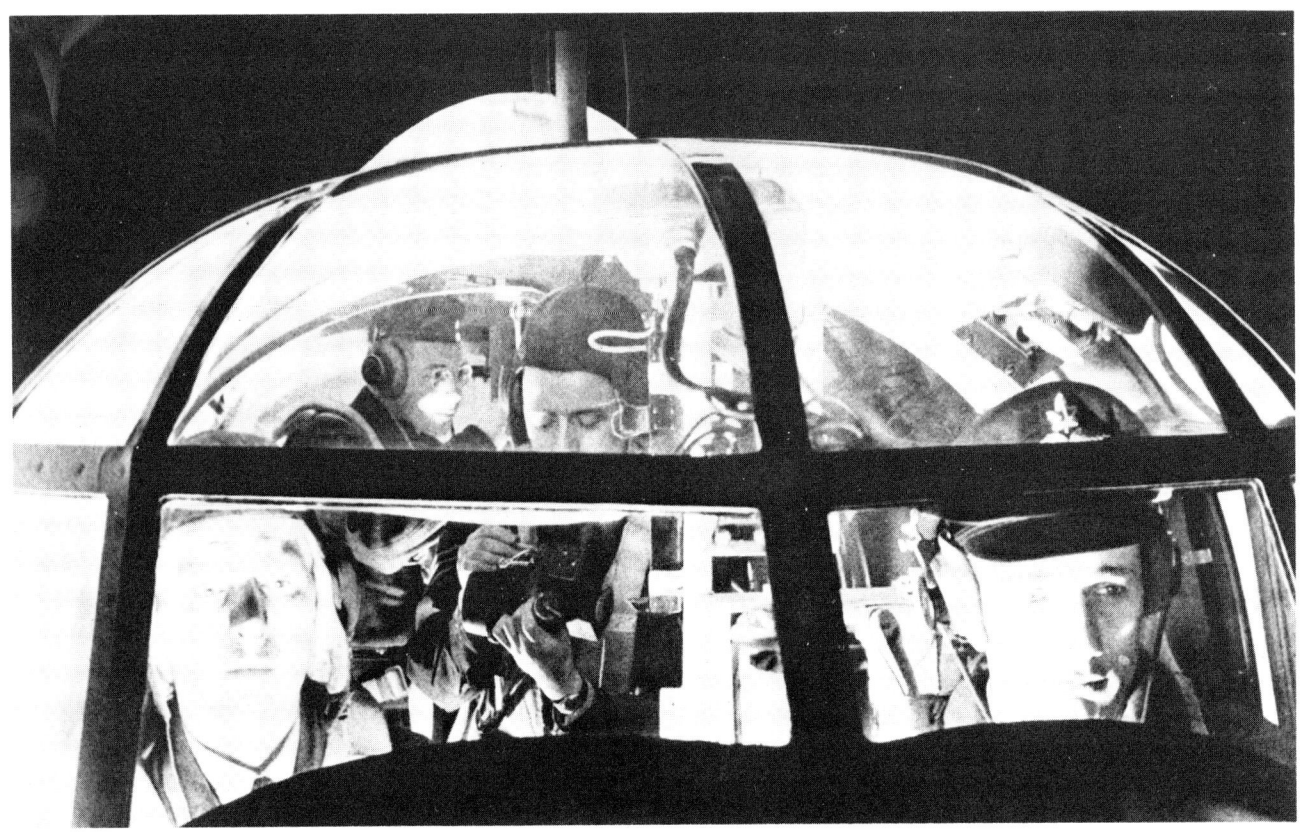

We'd go on navigational exercises and there'd be three or four of us in the Anson with the pilot. Somebody would have nature's call, and these aircraft were supplied with particular tubes and hoses to suit that occasion. When the chap would be ready to do his bit, we would signal the pilot and he'd pitch the control forward at the crucial moment. Some of those guys came back in a pretty sorry mess.

you'd say, "You're off course!" We had one chap who got five stitches in his head, and they wanted to know what kind of shenanigans he was doing. He just said, "I ran into a cloud."

Student Air Gunner, RCAF: Everybody got so bored with straightforward training exercises, and it must have been very monotonous for the pilots. They were all young guys and anxious to get into action. We weren't supposed to cross the US border, but quite often we would do a low level, and head down into Montana. It was real badlands down there—a lot like you see in the old western movies—and we had a lot of fun dodging in and out of the hills. Then there was Fort Peck Dam. I'd be riding in the mid-upper turret in the Bolingbroke and we'd do a simulated low level attack on this huge dam—it was very, very big—and we'd come in right down at ground level below it and just zoom up over the top.

Student Navigator, RCAF: I went for a ride in the co-pilot's seat in a Bolingbroke. The fellow who was piloting it had just completed a tour overseas and he was pretty pissed off about being shipped back to Canada on this sort of mundane thing, where two Ansons attack with camera guns and he tries to get out of the way. This guy was really cheesed off, so he started throwing the thing all over the sky just to get away from these Ansons. I don't even like going up in elevators, but I looked up, and there's the water up there, and then all of a sudden the sky is up there. He just threw that thing all over. At the end of the exercise the Ansons were flying side by side, wing to wing, and this guy went right in between. I'll bet there wasn't more than an inch and a half clearance. I knew if I looked at this son of a bitch he was going to have a big grin on his face, because he knew he'd just scared the hell outta me. Finally I couldn't stand it any longer and I turned, and there he was, big smile!

Student Navigator, RCAF: We'd be at the end of our exercise and two or three hundred miles from home, and this pilot would say, "Come on, climb in the seat." So we'd get high enough and he'd hold it, and slide out, and I'd slide in. He'd say, "Keep 'er straight and level." I always felt good. It was just between him and me because he knew I'd been washed out. And then I'd swing out and he'd take over again for the landing. He'd say, "Lemme take 'er in." He'd take her in, and nobody knew.

AOS Flight Supervisor, civilian: I'd go and get one of the Ansons from the maintenance hangar that had just been checked for something and take it up to test it, and that was always an opportunity to horse around a little bit. You always had two or three fellows riding with you—the student navigators—and they were all prone to want to horse around. So I did certain limited aerobatics and never had any problems. I looped them and I rolled them, and they'd spin right and left, and nothing mean happened. If you pulled them up in a stall and left a little power on, those old Ansons would more or less just sit there and shake, or you could put it into a shallow dive and pull back on the wheel, and it would go round in a circle just like any other aeroplane.

Instructor, RCAF: Quite a few instructors used to loop the Ansons and the odd one rolled them. But they took us down and let us have a look at one that had the wings stripped down. The Anson had a wooden box spar and thin metal ribs, and the spar was cracked and the ribs were twisted, because it had been rolled. They knew these things were happening, but they just couldn't catch anybody, and they wanted us to see what happened when you put an aircraft through stress that it wasn't built for. It kinda cured us.

Station CO, RCAF: The Anson wasn't built for anything more than straight and level flying, but these guys wouldn't listen, and every once in a while a spar would break and somebody lost their life. When the airframe mechanic came along and reported a cracked spar you'd go and check who was flying it last, but it's impossible to prove anything, and nothing ever came of it. We had Australian instructors for a while, and they were doing all sorts of things with those Ansons—rolls, loops, rolls off the top. They'd try anything, those darned Australians.

Instructor, RCAF: For a number of years I was both a counsellor and a camper at a boys' summer camp north of Montreal, and I had a real soft spot in my heart for this place. Don Coring was another instructor at Uplands, and one Sunday I thought, "I'll go down to the flight line and get hold of Don and see if he'd like to fly up to Stroder's Camp." He'd been a counsellor there too. He said, "Good idea!"

Very close to our track there was an Elementary School, and I had some friends there, so I told the pilot that I'd like to drop them a message. I put it in a "sick cup", and I was standing up. I'm not sure if he really went all the way over and did a roll, but I can remember looking out through the *top* of the aircraft and seeing the guys wave at me, and the sick cup rolled right down in front of the bunch of them. I guess he pulled a few Gs in that old Anson.

So we both took students and put them under the hood and set course for Stroder's Camp. We flew a loose formation on the way up, but when we got near we flipped the hoods and said, "OK, I have control," and we closed into a tight formation, put our noses down, and went right at the camp.

They have a waterfront there with a sort of pool, and their sailboats were lined up on the beach, and their canoes and their rowboats, and the clubhouse. And being Sunday, all the parents of the boys were up to visit. Then down across the lake came these two Harvards, very tight formation, howling like ding bats! On the deck and right at the beach, and at the last second we pulled up and zooommmed over the top of the clubhouse.

I guess we shook the timbers, because everybody scurried for cover, but we broke off our formation and then came down, and for fifteen minutes we put on a real show. We did rolls and loops, everything. It was a truly classic beat-up, and we reformed and made one final pass, and then set course back for Uplands. Then back at the mess we had a few chuckles about the great time we'd had!

Monday came—nothing happened. Tuesday . . . Wednesday . . . on Thursday morning my Flight Commander came to me and said, "John, you're wanted over at the CFI's office. Rob Portland's wanted, too." Rob and I were both warrant officers and both due our commissions, so we both let out a great war whoop because we figured, "Boy, our commissions!" So this was all that was on our minds as we walked over to the CFI's office—our commissions were going to be announced and we'd get all the congratulations that went with them.

The CFI's office was in the control tower building, and as Rob and I walked in one end who should come in the other end of the building—none other than Don Coring, my friend from the previous Sunday's excursion, with another warrant officer. I looked at Don, and Don looked at me, and immediately the light went on; our warrant officer chums were really our escorts!

I was called in first. I saluted smartly and quivered in my boots as the Wing Commander read out a charge against me for low flying. He read it out chapter and verse; whoever reported us certainly had it down! He described the whole thing in minute detail. He had the identification of the aeroplanes down to perfection. The Wing Commander never bothered to look up from the paper: "Mason, is this true?" And I said, "Yes, Sir." . . . "Then you're under close arrest." So off to my quarters I went, and Don Coring likewise.

I was called in to the Group Captain of the station. He said, "I could court martial you. We're looking to set an example, because there's been too much of this low flying going on." There was a bit of a silence, and he went on, "I'm not going to court martial you, but you're going to have your log book endorsed, you're going to be CB [confined to barracks] on the station for six weeks, and for every night of those six weeks you will be on night flying duty in the tower." So for six weeks, every night, I had to go up to the control tower and sit there and give the lights.

About ten days after I was finished my duty in the tower, I was called back in to see the Group Captain. He said, "Your commission has come through. . . . I could hold this up for at least six months." I didn't say anything. He said, "However . . . congratulations!"

Considering the mandate of the BCATP—to teach young men how to fly aeroplanes and the basics of aerial warfare—it is not surprising that there were accidents. What is *surprising is that there were so few accidents involving loss of life. The BCATP trained almost 138,000 airmen, most of them younger than today's university graduates, yet the records show that during all those hours of flying there were fewer than 900 fatal accidents that directly involved aeroplanes.*

Student Pilot, RCAF: There was a kid in my course who had been ill and was short on his flying time. He had to make up a few hours before he could get his Wings. So he was hustling around putting in time and went out to his family's farm and showed off, and cartwheeled around his dad's farmyard. They picked him up with a shovel.

Station CO, RCAF: At some points the casualties were pretty high. You'd get kids in there with a few hours and you'd turn them loose, and some of them weren't very level-headed, so they would do foolish things, and there were a lot of accidents. Quite a number of people killed themselves hitting trees and power lines and telephone poles, things like that.

Student Pilot, RAF: I was half-way through an instrument takeoff in a Harvard when the instructor grabbed the aeroplane, closed the throttle, and popped the hood just in time for me to see another Harvard go, *nnnyyyeooow . . . plonk!* right into the deck. It was a nasty smoking mess, and we came to a stop at the edge of the field. The old boy was out of there like a flash. He scrambled through the fence, got to the Harvard, took a look inside, and turned away. He came back and said, "You stay right there."

THE INSTRUCTOR GOT HIS CHIN CUT

Ground Staff, civilian: We had quite a few damaged aircraft. Sometimes both wings would be dragging on the ground, and there were lots of write-offs. One winter an American instructor had his student practising forced landings. It was a cold day, and he closed his throttle and glided down, and when he opened the throttle the engine was cold and there was nothing there. They were over heavy timber, and they went into it. The instructor got his chin cut, but he was very upset because the aeroplane was a total write-off, and we had to chop the trees down to haul it away.

The Flight Commander had been doing a Wings test, and somehow had allowed his pupil to stall just as he was making a circuit turn, and they lost sufficient height that there wasn't enough left to regain control, and they went straight in.

Student Navigator, RCAF: About a week before we graduated at Rivers we were doing night cross-countries. We saw this huge fire about three miles from the field, and we thought the farmers were burning straw stacks, but when we landed we found out that two aircraft had collided just prior to entering the circuit. That's my most vivid memory of Training Command—the loss of two aircraft and all the crew.

Student Pilot, RCAF: About eight of us were standing around outside our flight headquarters looking across the airfield, and an aircraft from another flight was in the process of taking off. It got about fifty feet in the air, and whoever was at the controls pulled back on the stick and the thing went straight up about 200 feet, and then it fell off and went over on its left side and went down into the ground. We all just stood there dumbfounded. Here's a group of us with three or four hours of dual, and apparently two people have just died on the other side of the airfield.

As it happened neither one of them was hurt aside from a few scratches, but our instructors got us into the aircraft and off the ground right away. That was standard procedure. Any time they had an accident they tried to get everybody up right away to get their mind off the incident.

Student Pilot, RCAF: This other Moth was landing at Fort William in the soft mud, and it went right over on its back. I thought, "God, he's killed." So I landed near him and ran over thinking, "Maybe I can get him out before it catches fire." I looked under, and he's hangin' in there by his straps, upside down. And he's cursin'—oh, he could curse! He's mad as hell: "Get me loose!" So I loosed the straps, but when I loosed them he came down and hit his head, and he's mad as hell at me! He's gonna knock the shit right outta me, because he landed on his head!

Instructor, RCAF: We had changed over to Cornells, and one morning I fought like a steer to get the aircraft I wanted. Everybody had their favourites, but I wasn't able to get it. The chap that got there ahead of me went off with his student, and another aircraft reported that he had observed the wings fall off, and both the instructor and student were killed.

This guy in a Tiger Moth had done a spin and was recovering, but he must have overcorrected. He went from a spin to the left to one to the right, and back and forth. Then somebody said, "Gees, he's not gonna get outta that." We watched him go right into the ground. He landed in a dry creek bed in the rocks—he couldn't have picked a worse spot. We went over to pick up the pieces, and he had a sprained ankle. What saved him was that he kept *the stick in his gut, because in a spin the aeroplane's not going very fast. And in a Tiger there's a lot to collapse up front before it gets to you.*

Instructor, RAF: All the Cornell stations started reporting strange incidents of crashes where the wings had come off. We thought, "It'll never happen here." People said, "The silly buggers, they're overstressing them." But we had a Canadian instructor come to the station. He had a bad cold and he was not supposed to fly above 2,000 feet, and he was in the circuit with a student when one wing came off. Of course, they were both killed.

They'd had fatalities at De Winton, and they'd had them at Bowden, so they blew the whistle: "Ground the Cornells. Inspect the main spar." So they drilled holes through the top of the wing and looked down, and sure as a gun the plywood reinforcements were splitting away from the main parts of the spar at the glue line.

Instructor, RAF: They had to haul them all in and check them and reglue the plywood. A lot of people got the wind up, and some of the pupils wanted to get off the course, and some of the instructors that were married wouldn't fly. "We've got to have a clean bill of health."

Instructor, RAF: No solo student killed himself while I was at Assiniboia. When I was at Bowden we had two students killed, but they were dual, and we had no solo students ever kill themselves there. It was usually split arse instructors playing silly bugger—low flying and zooming around and, bang! They're both in, and both killed.

Ground Staff, civilian: We had an instructor with a student that were entertaining people at lakefront cottages by low flying, and they struck a tree, and that ended both careers right there.

Ground Staff, RAF: There was one pilot who had the habit of going over one of the lakes with a Fairey Battle and circling low and making ripples on the water with his wing tip. But one time he misjudged, and in she went.

Ground Staff, civilian: In one accident with a Fleet both the instructor and the student were killed. They were doing loops, and they did three in a row, and they didn't pull out of the last one. They were both killed, and the thing caught fire and burned. We found the tail wheel.

Instructor, RCAF: We had three fatal accidents on Fridays, and it so happened that they were right after lunch, three weeks in a row. On the fourth Friday there were fifty students on sick parade!

Ground Staff, RAF: One time we lost nine aircraft around Medicine Hat, all in one morning when the weather clamped in and the student pilots and instructors got lost. It was wintertime, and we needed nine crash crews, because every aircraft was in a different place in the foothills.

We found one aircraft with the wheels up where he had belly landed. The instructor and student were both dead. They were frozen in their seats, and we had a hell of a time getting them out. And they found a couple of the aircraft that had bumped into a hillside. It took us five days to find all nine.

Student Pilot, RAF: We were doing a low level cross-country, flying at fifty feet over the prairie. I was on a compass course, but all of a sudden I caught sight of something on my right—a great

The Cornell was a funny-looking low-wing thing, and the wings used to fall off any old time when you were doing aerobatics. They had a red line not to go over 120 mph, and of course you weren't allowed to do flick rolls. As far as pulling out too fast from a loop... the wings just folded; the spar broke. Quite frankly, we were scared to fly the damned things.

burst of orange flame—and I called to my instructor, "There's something odd over there." He took control and turned, and we looked and saw two aircraft blazing on the ground. It transpired that two fellows on our course were flying low level and collided—an awful mess. I'll never forget that colour; it wasn't like an oil lamp, it was more of an orange colour, and at first the flames seemed to bounce. I suppose it was the broken parts of the aircraft that caught my eye as they bounced on the ground. It was a terrible sight.

Student Pilot, RCAF: We went in and landed just to the west of MacLeod beside the remains of what had been the missing Tiger Moth. There was nothing but a pile of white ash where the engine had been, and the remains of the men. We didn't touch a thing; we just came back and reported it. Then I had to go back out there and stand watch all night, and the groundcrew came out next morning to pick up what was left. That wasn't a very pleasant experience. Those guys were burned clean. There's one thing about the old Tiger: when she went on fire, she *really* went on fire, because of the way the gas tank was held on the struts. If there was a heavy impact, those struts went right through the tank.

Student Navigator, RAAF: We got word that a Fairey Battle had crashed and the pilot had been killed and the two Australian students had been killed. Up till then it had all been fun and glory sort of thing. you were in the Air Force; you had your uniform and so on. And then it sort of hit home. "Boy, you could get *killed* in this game."

They gave them a very nice funeral, and the whole station turned out. They went to a place called Wynward and they took us out there in buses. It was a cold day, but they marched the group right out to the grave and they fired volleys over the two coffins as they were lowered. They had the Union Jacks over the coffins. Afterwards back at the station, all the boys sort of had a party. They said, "If they'd been here this is what they would have wanted. They didn't want any gloom and sorrow, so let's remember them, and remember them in fun." And they did, but it was done very nicely.

Ground Staff, civilian: It was a solo student that went in, and it was a demoralizing sort of experience. You never think of these things happening, but it did, so there was really quite a ceremony. It was the first on the station, and at the beginning there was a line of thinking going that it was a mechanical failure, and this had a very bad effect on us maintenance people. Those aeroplanes were *always* tested after any major work was done.

Airman, RCAF: We all attended the funeral, which was a very solemn affair. All the students were in dress uniform, and we marched out to the local cemetery, which seemed just like any other small town cemetery. Everything else at the station stopped for the one day.

Airman, RCAF: I couldn't have been on any more than five funeral parades. A lot of guys were getting killed in Training Command. We didn't know them necessarily, but I was either on the firing party or the escort party.

The casket was put on a flatbed RCAF truck that had a big green mat on it. The casket went up on the mat and then the flag went over it. The firing party were all lined up, and when the casket came out to the grave site we'd go to the "present arms", and we'd put this poor chappie in the ground. Having not known the person and just being part of the ritual, I really didn't like doing it, but I got stuck for it.

At some of those funerals we left the casket on the ground, but at others they let go of the straps and we lowered it down. There was the flag, and the airman's hat with the little white flash, but they never let that go down to the bottom. Somebody was detailed to take the flag and the hat off before the thing disappeared. I always thought that was cheap in a way, but I guess it was drill.

Airman, RAF: There was a young boy, Harry Hall, eighteen. A bright, cheerie little boy. He used to say his prayers on his knees at night, and in a hut of fifty guys this was a brave thing to do. He had an instructor who was a DFC, and they went out to the relief field at Airdrie and Harry went solo. The instructor climbed back in and they took off, and they had an engine failure, and the thing spun in and they were both killed. An awful tragedy.

I wrote to his folks and I told the Flight Sergeant that I would like to be one of the pallbearers. When we got down to the grave there was no cap on the coffin. So the Flight Sergeant took my cap and put it on the coffin. As they were lowering it into the grave, he took the cap back and handed it to me, and I said, "I don't want it." He said, "Take that cap!" I said, "Flight Sergeant, I don't want it, now." . . . "Are you gonna march into town without your cap?" I said, "On this occasion, Flight Sergeant, I think I will." I never wore it again; I just felt odd about that.

I'D GIVE THEM ALL A GOOD SHOT OR TWO

Ground Staff, RCAF: They'd give me a crash crew of four men, and if it was winter we'd hire a snowmobile. They were homemade affairs, not like the snowmobiles we have today; some of the farmers built their own. I'd get to town and buy a bottle of scotch or rye, or *any* kind of liquor I could get hold of, then before we'd go down to the crash I'd give them all a good shot or two. It'd be awful—terrible gory mess. Sometimes the head would be in the rear cockpit. And when there were two aircraft involved, like a mid-air collision, you really couldn't tell who was who. I used to put a blanket on the ground and just divide up the meat sort of thing—put a label on it and send the package back with the snowmobile.

Despite the high degree of organization that went into the efficient operation of the Commonwealth Training Plan, it was inevitable that mistakes be made. Sometimes these were brought about by the course of the war, sometimes by government policy, and occasionally by ignorance. When individuals were affected by moves that might now appear thoughtless or stupid, they were generally brought about for reasons which then seemed important. Occasionally, an apparently stupid situation arose through circumstances over which no one had control.

Student Pilot, RCAF: We were posted to Elementary Flying School and we went to St. Eugene by mistake. It was in the late winter, and they hadn't attempted to remove the snow, they just compacted it. Once that started to melt it was game over, because you'd get stuck in the slush. So there was a period of about a month when they couldn't do any training, and that was the time when we arrived at St. Eugene. They had two courses sitting around waiting, but the field was broken up and they couldn't handle us, so we were put back on the train and shipped off to Goderich.

Washed-out Pilot, RCAF: They sent me to Mountain View, and I was there from April of '44 until August, and I did absolutely nothing. The station was filled with washed-out pilots who were waiting to go on navigator/bomber courses, and Mountain View was for the bombaiming and gunnery part of the course. There was an officer there in charge of us washed-out guys, and it was up to him to see that you got on a course. Every two weeks a contingent of trainees would arrive, and there had to be so many, and if there was room they would take two or three of us and put us on to fill it out. It was all entirely up to this one officer.

Instructor, RCAF: They had one aircraft they called the Fleet Fort, and I refused to fly that. At 4,000 feet above sea level it was overloaded, and they were bursting into flame on takeoff and things like that. I caused a regular little . . . it would have been mutiny had I lost, but Western Command engineering people decided it would be best that they be grounded.

Instructor, RCAF: Those Forts were terrible aeroplanes. It looked a bit like a Harvard, only smaller, and I think it had a Jacobs engine of about 300 horsepower. They used to develop fatigue in the tail, and the gas caps used to come off and siphon all the fuel out. Worst of all, they had multiple disc brakes, and if you used them too much before takeoff they'd get hot and the discs would warp. Going down the runway they'd get white hot, and once you were in the air they'd weld, and the wheel would never turn again. On landing you'd go right over on your back. They were murderous things.

Instructor, RCAF: It was very strange in the Air Force. Somebody would come to you and say, "You gotta be checked out." So I was checked out by the Chief Flying Instructor, who was RCAF. I don't know if he was trying to impress me, but we went through the whole rigmarole of what-to-do-in-a-Finch. I had just been posted to this EFTS from CFS Trenton, and I had previously taken my own Elementary on Finches, but this CFI insisted on going through everything, including aerobatics and low flying. We were over the low flying field and he was demonstrating how to do a slipping turn properly. He came in with a great rate of descent and pulled it out straight, but then we hit the ground with the two mains, bounced about ten feet into the air, and just kept right on going. That was my introduction to being refreshed on the Fleet Finch.

Student Pilot, RCAF: I've often wondered about this business of taking someone who pooched and just washing him out. My buddy hit a tree. OK, we were low flying, but we were nineteen years old and flying a twin-engine aeroplane that perhaps we had no business in anyway. He and I had done quite a bit of flying together, and from the flying point of view he was right up at the top of the class. But they sent him back to England, having already spent all that money training him. Then he spent the rest of the war doing some erk job, when he really should have been flying aeroplanes.

In a way they were too quick to jump to the conclusion, "The guy pooched. Get rid of him." But many times things happen to you in an aeroplane, and that's really what you learn from.

Student Pilot, RCAF: The CFI assigned me to an instructor by the name of Bernie, a civilian, and Bernie was not my type, or I wasn't his type, whichever way you want to look at it. We didn't get along very well. He was a very dirty-mouthed individual, and I couldn't stand being sworn at; it was just too much for me to take. All my chums went up for their initial familiarization and they had fun doing a few aerobatics and stuff, but I went with Bernie, and, "You're gonna start learnin' t'fly right away!"

205

We used to have to go out to do emergency landings, and there was a little wee wind sock on a fence post. He said, "You see that sock down there?" And I looked, but I couldn't see any wind sock. He made me turn: "Do you see it now?" And I said, "No." I really didn't know what I was looking for; it wasn't coloured, it was white, and it was about a foot and a half long, and it was on a four-foot fence post. I never did see it. Finally he just got so mad that he grabbed the stick out of my hands and flew down to about two feet off the ground, and just about ran into the fence in a rage: "Ya blind son of a bitch! D'ya see it *now*?" And by this time I'm so upset I couldn't even see the inside of the cockpit.

Officer, RAF: Training Command Headquarters had some of the dumbest people I've ever come across—completely inexperienced, but because they were Command Headquarters, they knew it all.

A guy got killed. He took off at night thinking his gyro horizon was nice and level, and he flew on and on and it was still nice and level when he went upside down. This guy from HQ came out to the station and decided what had happened, before he had even started his investigation. H'd never flown himself, but he set out to collect any evidence that might help him prove his theory. He was trying to prove that this pilot took off looking over his shoulder at the lights behind him, and lost control and spun in. It turned out that that type of gyro could be caged, and there were quite a few cases where students took off with caged gyros, with varying degrees of disaster.

Instructor, RCAF: If the Air Force got what they thought was a good instructor they hung on to him, but instructors burned out and Training Command couldn't believe this. For the first three years of the war they thought there was no such thing as flying fatigue for an instructor; they just didn't believe in it. Later on if they even sniffed it, they'd send an instructor on another posting for six months. But it wasn't until well on in the war that they decided that a term of instruction would be a year and a half.

Senior Officer, RCAF: We were forever asking them to let us know what was necessary overseas, but there was no feedback. We said, "Can't you send expired operational pilots who have done a tour to tell us what is required?" We were teaching pretty tight formation flying, but we found out eventually that the bombers that were going on ops were using a much looser formation. And they were flying at night, but nobody ever told us we should practise *night* formation flying.

WAG, RCAF: To graduate I had to learn Morse at 32 words a minute, but when I went to operations, eight words a minute was more realistic. I thought, "What a stupid thing—all the poor guys that washed out because they couldn't get up to speed."

When we arrived in England we found the radio equipment we'd been training on was all obsolete, so it meant retraining practically all over again. And we had very little training in radar, which was all the thing in England. I don't even remember radar being mentioned when we were in Canada; it was top secret and there was no talk about it at all.

They would have saved a lot of grief if they had trained crews in Canada with the latest equipment and the latest aircraft. It would then have taken very little to get them into ops. But I spent a good nine months just retraining once I got to England.

Pilot, RCAF: It was difficult to understand the way they handled the Wings graduation. Those that stayed back in Canada as instructors, every one of them got their commissions, and every one of us that went overseas in *my* draft went over as NCOs. Yet when I got into the squadron most of my crew was a higher rank than I was. They had just graduated earlier and been waiting, or they'd been off somewhere else, and a washed-out pilot would wind up as a bombaimer or a navigator with a commission. That was something that was very disturbing to us pilots who were put in charge of the crew. If a decision was poor, it was always the pilot that got hell for it, and that was disturbing.

In spite of the gaffes and the bumblings, in spite of the wash outs and the apparently unreasonable postings, in spite of the weather and the mud and the cold, there was that final posting overseas that everyone craved. For many, the route there turned out to be the most unpredictable leg of the whole training program.

There was a fellow in my home town by the name of Pete Murphy. We had to report to Halifax at the same time. Now, Pete liked women and he thought he could drink. I don't know what kind of a lover he was, but it didn't take too many drinks to get Pete pretty sloshed.

Montreal was the big depot for heading for Halifax, and in those days you came in from the west to one station, then changed over to another station to continue on east. Our train didn't leave till eleven o'clock, so I called up my grandmother and took her to dinner, just to show her my big three hooks and my Wings. Pete was supposed to meet me a little before eleven and we'd get on the train together. Well, he missed the train, and I thought, "Gees, just goin' from one station to the

Station CO, RCAF: *We used to turn out about forty pilots every three weeks, and some were selected for commissions, and some were selected as sergeant pilots, by guess and by God. It was based somewhat on their ability of flying, but I never did quite know how they did that; I never was quite sure there was a fair system. There were just so many officers and so many sergeants, and I'm not sure how the hell they decided. It didn't work very well, because when they got into crews in the bombers, half of them were non-commissioned and the rest were commissioned, and it was often a very difficult situation.*

other Pete got picked up, but he'll be along on the next train."

We arrived the next day at Y Depot in Halifax. There were *carloads* of airmen there, sergeants and officers, all aircrew to go overseas. They lined everybody up in a hurry and called the roll, and when they came to Murphy—"Murphy, Sergeant Peter"—nobody was around, and I yelled "Present!" 'cause I figured he'd be in on the next train. So we went and we ate. Then next morning we did the same thing, and in the afternoon again. So now I'm well into the second day and Murphy hasn't shown up yet, and rumour has it that the ship we're going on is pretty close.

Well, I chickened out and quit answering "Present" for him, so suddenly Pete's absent. We had a Warrant Officer there who was one of these fellows who went back and forth with the drafts and he'd never lost anybody till this guy Murphy, and he was gonna find Murphy before we went. Going up the gangplank there was this Warrant Officer asking everybody, "Have you seen Murphy? . . . Have you seen Murphy? . . . Have you seen Murphy?" I thought, "Boy, Pete's really going to get it. He's going to be marked AWOL from an overseas draft," which is a pretty serious offense. So that's where I just sort of forgot the whole thing.

I was in southern England about a week and I looked up one foggy night, and who's walking along but Pete Murphy. He'd come over on the next ship, and he said, "You know, I was loose for quite a while. I was shacked up with some babe, but my money ran out, and was I ever in trouble when I got to Halifax. They tried to tell me I'd been there and *then* gone loose." So I told him, and we had a big laugh.

The train took us from Toronto to Montreal. We got into Montreal around noon and we had to connect with the Atlantic Limited at night, so we had the whole afternoon off. We went up Mount Royal and we had a ride in a caleche and we sort of did St. Catherine Street.

There were five of us and we'd been issued meal tickets for the Windsor, so we went into the Windsor Hotel, and it was palatial. There was an eight-foot *maitre d'* with a real Gaulic nose that met us at the entrance to the dining room. He sort of looked down at us, but led us to a table, and they brought us a menu as big as the Free Press. They had lamb chops, and I said, "Do they have those little frilly papers on the ends?" And he assured me they did. So that's what I wanted, because I'd never eaten lamb chops in that fashion.

We had a fantastic meal, and finally the waiter brought the bill and the five of us pulled out our meal chits. He looked at them aghast and disappeared, then came back with this eight-foot *maitre d'*, and *he* said, "These chits are for the Windsor *Station.*" So we sort of looked at him and said, "Oh." He said, "Where are you going?" . . . "Halifax." He said, "Have a good war!" and that was the end of it.

We met the rest of the gang at the railway station, and they were really bitching that the cheese sandwiches that they got in the cafeteria were two days old. We said, "We had a delightful dinner." There's a lot of innocence in being young!

We were posted to Y Depot in Halifax, and I never did find out what a Y Depot was. Halifax didn't impress me too much. I guess it wasn't the best place to be in wartime and there wasn't too much to do—you just waited.

There was a real air of war. You saw the Navy all around. You saw the Naval Ratings in the streets, and the ships coming and going, and tremendous convoys forming up and being escorted out to sea.

A lot of the boys had come right off the prairies out of central Canada and they had never been near the coast; they'd never seen any port. So it was very exciting for them.

We didn't spend too much time in Halifax. I was kinda glad, because it was a dingy, crummy place during wartime. It was mainly Naval personnel there, and there wasn't too much for an airman to do, so when we were loaded on the ship it was a relief to get away. After Halifax anything would look good.

We were told not to talk about anything, but we were leaving, and by God, if we didn't put our kit bags over our shoulders and march from Y Depot to the ship, the *Aquitania*. Bands were playing, and we were not supposed to tell the Germans we were coming, but we were advertising! People on the sidewalks, waving: "Another draft going overseas!" There was only the one ship in the harbour, the *Aquitania,* and they all knew where we were going. And here in Canada we're getting all this, "A slip of the lip will sink the ship," and big posters of Hitler, peeking around the corner and listening with a big ear!

The train went pretty well right down to the wharf, and there was a ship standing at the dock; it was

WE DON'T STOP FOR ANYBODY

On the way over on the Queen E there were a few submarine warnings and all kinds of rumours. Somebody said there were about 20,000 men on that ship. There sure were a lot of Americans, because it loaded up in New York first and then came into Halifax and picked up the Canadians. So the majority were US Army and the rest were all RCAF.

There were the usual stories from the civilian crew: "We don't stop for anybody," and "Somebody dropped overboard last night and we couldn't go back for him."

Off the coast of Ireland a couple of Sunderlands came out and circled, and as we got closer we saw other Maritime Command aircraft, and the odd fighter would go by. My greatest excitement was when I saw my first Spitfire.

the *Stratheden*. I knew the *Stratheden* was a P & O ship that used to travel from England to Australia. It was a lovely ship, and I said, "Oh, the *Stratheden*, this'll be good."

There was a fellow to show us the way, and we went away down into the hold. They had hammocks hung up, so we weren't going to have the normal passenger accommodation. They had brought a bunch of war prisoners from South Africa to Halifax, and the ship was a mess—it was filthy. The fellows got together and said, "We're not gonna stand for this!" So they went on deck and found the chap that was in charge of all the troops on the ship and said, "This is filthy. Nobody should be asked to travel on this." He came down and said, "Yes, things aren't too good. I'll tell you what. Just make do tonight, and tomorrow we'll get things cleaned up and we'll make some other arrangements." So we went downstairs and we slept that night. But the stinkers—they moved the boat away out into the middle of the harbour somewhere, and there was no chance at all of getting off. So we stayed there.

It was a funny thing, as soon as we cleared the harbour at Halifax, we hit those big waves. I happened to be out on deck and it looked like we were climbing a hill to get out of them—Jesus, they were high. I thought the ocean would be level, but then they started splashing over the deck, and when night fell all the decks were blacked out. And my stomach started to heave.

I tried to sleep in the hammock, but I couldn't sleep in those buggers; I kept falling out. I tried sleeping on mess tables—that wasn't too good. Finally I spent the nights sleeping on stairs between decks. That's the only place I found where I could get any sleep at all. When I look back on it, it's a good thing the Navy didn't get me before the Air Force.

Where we were quartered there were wooden bunks, four in a row, and I had a lower with three other guys above me. Being inside a pitching place like that and everything blacked out, it was no wonder you were bloody well sick.

It was terrible—simply awful. I slept underneath a mess table for all the time I was there. I started out in a hammock the first night, and fell out of that and took about four guys with me. And after that I just stayed under the mess table.

That P & O boat was built for the South Pacific run; it wasn't built for the north Atlantic. I can remember the hammock hitting the underside of the deck above us, and then all of a sudden whipping back and hitting the other side. And there was a terrible torrent of water came down the stairways and into the hold. The ship was just flooded, and I thought we'd turned over.

The next day the hold was full. All the garbage cans had come down, the piano had broken loose,

We were supposed to sleep in our clothes in our hammocks, but after a couple of nights I was just not sleeping that way. But those hammocks and blankets had crossed the Atlantic many times with many different people, so there were a couple of dozen of us that started to scratch in unusual places. When we got off we had added something to our beings in the form of crabs . . . and that was our introduction to that aspect of our careers!

and the galley was a hopeless mess, so there was a big clean-up day. The following day we had everything stashed away before lunch and we were sitting down to eat at the long mess tables. They brought in big tins of something like corned beef and cabbage—not the most appetizing food—and the tins were dumped on the tables. Then the ship started to roll. The deck was probably as clean as it had ever been, and we tried to grab these big tins of food—they were about two feet by three feet—and they started sliding down the table, and slid back, and we watched—back and forth, and finally off the end. God Almighty! The next thing, sea water came rolling down the deck.

Any time they called "Quarters" it was a lifeboat drill, and you had to rush out on deck to be counted. And, Jesus Christ, there was always an officer there with a sub-machine gun. I wondered what he was going to do to us.

They needed volunteers for gun crews to serve on the various watches. It must have been about a three-inch cannon that was stuck out on the stern, and they were looking for crews. I volunteered because when you came off watch you got fed, which meant you got three meals a day instead of two. You got a cup of cocoa and you got a cheese sandwich with the bread cut about an inch and a half thick, and a big wedge of cheese, and that was another meal. During your watch you stood out on the butt end of the ship and watched and waited. I don't know what would have happened if we had ever had to use the gun; I'm sure we'd all have been paralyzed with fright.

They used to feed us twice a day, once at six o'clock in the morning and once at six o'clock at night. If you wanted to line up for the little nappies or whatever they had to sell, you were in about a four-hour line. A long wait just to get a package of peanuts.

You got in line in the morning and by the time you got up there it was time to eat, and if you went right back to the end of the line it was time to eat again. You had your white cards and your blue cards with you, but this thing wound right around the ship. Of course, there were *some* fellows that *never* went.

Breakfast was powdered eggs, and lots of Spam for dinner. God, it seems to me it was Spam for dinner and Spam for breakfast—Spam all the time. The long tin-covered tables must have had twenty people on each side. There were people designated to serve each table, and depending on which end they served first, you either got a good selection, or you got what was left when the tray got down to the other end.

Pretty well every morning for breakfast we had kippers, and if anybody feels a little bit seasick, there's nothing worse than kippers. I'd go down to the mess hall, but everything smelled repulsive. Whatever the hell it was I couldn't eat their food. I'd swallow a chunk of bread now and again. Luckily, I'd brought a lot of bananas and oranges and apples and grapes, all that kind of fruit, a whole bagful, and that's what I ate on the way over. I couldn't stomach their food.

The ship was never designed to take all the troops. The washrooms weren't flushing. Gees, you'd open the door and you'd walk into the washroom and you'd step into about three inches of . . . water. All the toilets were either overflowing or flushing back.

When you went for a leak you were going to be at least up to your ankles. This stuff was up to the bulkhead doors. Nothing was working. They'd overcrowded this ship with thousands of men, and this water was always sloshing around.

The whole bloody place just stank . . . of vomit, and urine, and everything. Talk about being sick! You'd puke so bloody much. I rushed into the can one night thinking, "Well, there's the can; just hold on here; and let go!" And I held on and let go . . . right into somebody's lap. I had no idea there was anybody there. He got up swearing and cursing and swinging, and I got back into where the bunks were and of course he couldn't find me. But I'm sure he was gonna pound the shit right outta me.

It was the first trip the *Amsterdam* had made as a troop carrier. Everything was in Dutch on that ship. There were no English signs or anything. I was sick as a dog! God! I puked and I did everything. For four days I wanted to die. Then I looked through a port and I saw a rope. We hadn't even moved out of Halifax! And there I was . . . sick . . . oh, gees, it was all in my head. Oh, God, was I sick, and still tied to the dock.

We arrived at Greenock in Scotland. That was really exciting, because we were in a war zone, and as we were coming in there were all kinds of British submarines and warships, destroyers and aircraft carriers, and all kinds of Allied aircraft flying about. You suddenly realized, "This is for keeps; this is the real thing."

Bournemouth was our first taste of Britain, and it was pretty nice waiting there for postings to Operational Training Units. Lots of girls; we just had a ball in Bournmouth.

It was a fantastic place, just like being checked into the Empress Hotel in Victoria or the Royal York in Toronto. The very best of everything was right there for you.

We were bundled into a train and transported right down the whole length of England to Bournemouth. It was a gorgeous old car with lovely upholstery, but they had vertical backed seats and they were made for fifteen or twenty-mile runs into the cities from outlying communities. We kind of laughed at the little trains, because they were so small compared to our Canadian ones. Sometimes they were going very slowly, but at other times they were going through stations like pickets on a fence. It was just incredible the speed those things travelled at times, and all my teeth felt as if they were coming loose.

We arrived at Bournemouth, and after a cold and rough passage it was kinda nice to walk around in the sunshine with our shirt sleeves rolled up. This was in February, and it was really nice and warm in the southern part of England. Bournemouth is a real resort city with lovely hotels, and all of these had been commandeered to house Canadian airmen. This was the disembarkation depot for aircrew coming from Canada.

It was one of the main resorts in southern England, right on the coast. The Air Force had taken over quite a number of hotels, but we were billetted in a big apartment block. All we did in Bournemouth was parade in the morning, more or less a roll call, and after that we were free till the next morning.

The girls from all over England used to come to Bournemouth for their holidays, so there was no shortage of girls.

You could either go swimming or play golf, and they did have a few little route marches for exercise. We used to choose golf, but we never did get on the golf course because they had a bar, so we'd spend the day there.

There was an immense number of RCAF people there, and enthusiasm was heightened because people would come along and beat up the beaches with P-38s. Typhoons, Beaufighters, Spitfires, Hurricanes, anything you could imagine went whistling by there.

It was a real military zone. The FW 190s used to come right up the channel and strafe and drop bombs. Just before we arrived one of the converted hotels holding Australian airmen had received a direct hit, and there were all kinds of them killed.

I experienced my first air raid in Bournemouth, and that day there were 144 people killed. It was a Sunday, the 24th of May, and three Focke Wulfs came over and strafed the park. I guess they knew that at noon hour it was filled with aircrew. They asked for volunteers, and there was nothing else to do, so for the rest of the week I dug graves.

Everybody had a bicycle. That's the first thing you did, as soon as you got to England you'd buy a bicycle, because it was the only way to get around. A lot of people didn't take their bikes with them when they were posted and they'd sell them right on the spot. Or they'd be broke, so they'd sell their bike, and you could pick them up for eight or ten dollars. So we went riding around on our bikes waiting for pay day, visiting the pubs and playing pool.

A lot of us thought we knew the English money system, but really we didn't know a damn thing. When we went to the pubs for a pint of bitters we'd just hold out our hands and the little old barmaids would take whatever they wanted, and we didn't know how much we were spending. It sure makes you learn how to count money in a hell of a hurry.

I was freezing to death in Bournemouth. I wore my winter underwear all the time, because I wasn't used to the English climate in January. And I just roamed the streets. It may be beautiful in summer, but at that time of the year it's strictly from hunger.

Lord Haw Haw was a very popular broadcaster. More people in England listened to him than anybody else, and he was always kidding about Bournemouth: "We know about you Canadians over there, and by the way, your clock is about five minutes slow." And there was a big town clock, and sure enough, he was always right!

Haw Haw was the British fellow that had gone over to Germany, and he used to come on the radio right after the BBC news at nine o'clock. One time he said, "There's a lot of Canadians just arrived in Bournemouth," and he even named the ship that we had come over on. He said, "If you don't believe that I know what's going on in Bournemouth, then look up at the clock on the hotel where you have lunch. You'll notice that it's two minutes slow." So that kinda shook some fellows; they wondered what kind of a spy system the Germans had going.

Lady Ryder had a tea dance for us every day at four o'clock in the afternoon. She arranged hostesses and so on and we could go there and have a cup of tea and crumpets. It was very well attended. This was something that we Canadians had never had before, and we were there in droves. She felt she was doing something for us, and indeed she was, and it was appreciated.

Contact flying in England is quite different. The lovely Canadian section lines—north-south and east-west—give way to a can of spaghetti. With

I had had far too much to drink and nearly gave up my life by heroically falling off a bus. I was saying goodnight to a cute little girl in the crowd and then I ran and tried to jump on a bus. I don't know how I did it, but I fell off the bus and never stopped moving. I kinda cartwheeled over, ran around the back into the crowd, found her, and escorted her home. When I got back to the mess the next morning I had imprints of the hub of a wheel on the back of my jacket. You could see the greasy stud marks.

confusing rail lines and heavy smog in the industrial areas, you really had a tough time. To a Canadian pilot used to the West, to take off in misty England and turn downwind and then, "Gawd! Where's the field?" it's a funny feeling.

One of the first things I did was to get lost in a Tiger Moth. I looked down and there was a railroad track. I thought, "I should be able to find that on the map." I did, but there were eleven others exactly like it. So I ended up at a US Air Force base where they were doing glider takeoff practice with Dakotas. Just as I turned final there was a Dak came in and dropped two of these ruddy gliders right in front of me.

I managed to get the thing down, and there's a jeep with a "Follow Me" sign on it and he takes off down the taxiway at about sixty miles an hour. Of course, the Tiger can't taxi that fast, but I followed him to the control tower, and this big, fat American with his cigar comes out and says, "What the hell are you doin' on maa ai-drome?"

Familiarization with new terrain and different navigational hazards was only the beginning of a whole new phase of training for the airmen from Canada. The British had accepted responsibility for seeing that these men became competent on front-line fighters and on single and multi-engine bombers. This required postings to a stream of additional stations where a variety of conversion exercises were carried out and where extra training on new types of equipment and operational procedures readied them for their dreamed-of final postings to an operational squadron.

Ops required more than familiarization with new and much heavier equipment. The operational aircraft were certainly far different from the Harvards and Ansons and Fairey Battles that these young airmen had become so accustomed to, but now another element was added to their training: working with other airmen. The bombers could not be handled by a pilot alone, and the formation of efficient and effective teams became all important.

Pilot: They put an exact number of pilots, WAGs, gunners, navigators, and so on together. You were put in one big room and told, "Go pick a crew!" You just wandered around and you would see an observer and you would talk to him, and then you would talk to a gunner, and finally you built up your group. My choice for navigator was from Saskatchewan, and I said, "What does your father do?" And he said, "Well, he's the town bootlegger." So I said, "That's fine. You're my navigator." I figured anyone who was the son of a fine gentleman like that would be good. My rear gunner said, " . . . from Kingston." I said, "Oh, RMC?" He said, "No, penitentiary." And I figured, "Here's another good one," so I took him as my rear gunner, and indeed they all worked out beautifully. Somehow you just got together.

Pilot: You're all on the same station, but you're not all in the same mess, because at that time I was still a sergeant pilot. Somebody told me that these two fellows were sticking together and that they wanted to be picked together. They were both officers, but because it was the pilot's responsibility *I* had to approach *them,* and ask them if they would fly with me. I had to screw up my courage to do this, because what can a sergeant say to these people who are officers?

Officer, RCAF: The pilot was the crew master, and the navigator and the bombaimer and so on depended on him for their life. So as long as he was a good pilot, I don't think there was much of a problem when it came to ops. If the pilot wasn't good then the crew wasn't very happy, no matter what his rank was. But on operations the pilot would have the say.

Navigator: I didn't want to fly with any English crew, I wanted to fly with Canadians. But where the crewing up was, they were all Australians. I thought, "Oh, my gosh, that's all I need. To go with these mad Australians is the kiss of death." They didn't care for anybody else; they were individuals. Their term for endearment is "a dirty rotten bastard", and a bastard to me is one of the lowest things. But that was what they called you, and I learned to use that term myself, in its affectionate sort of way.

I crewed up with three Australians, and from then on *I* was an Australian. I did my tour with an Australian squadron; I wore an Australian uniform; I put an Australian Wing up; and I had the biggest mustache ever! If I kept my mouth shut, they thought I was an Australian.

WAG: I remember filling out a questionnaire, and we were all very excited because, boy, they were giving us a choice. "What war theatre would you like to serve in? What type of aircraft?" I thought, "Boy, this England is too cold for me, I'd like to serve in the Mediterranean somewhere, and it'd be nice to fly Bostons or Mitchells." So I put that down. But I think it was an exercise you went through. I ended up on heavy bombers in England.

Pilot: I used to have to pee a lot. I'd pee on the ground by the aircraft before we took off and maybe two or three times before that in briefing. Then I'd get into the aircraft and I'd have to hold myself till we got up to 10,000 feet, which took about forty minutes in those old birds. By the time I got to height, gees, I was just in *pain*. But I still wanted to get the aircraft on course and get the engines all synchronized and so on, and then I'd reach down and pull this little tube out and relieve myself. On the bombers we had a little outlet on the left side of the aircraft that faced backwards; then you had a tube, and as you'd pee it just went out in the slipstream.

Well, this day our aircraft had just been on inspection, so we'd got it out of the hangar and we'd got up to height, and again I was in this terrible pain, but I had my engines all synchronized so I reached down. And you have to get your little dooey there, and it's cold. You go through your parachute harness; you go through your flying suit; you go through your battle dress; you go through your shorts; and I used to wear pyjamas too, just in case I was ever shot down—I thought they might keep me a little warmer. When you get to the damn thing it's about as long as the end of your thumb, and you have to get it out.

So I finally get it out, but I have to go so bad that once I start I can't stop, and it's *pouring* out. And then all of sudden it all comes back in my face! I couldn't stop; I had to go so bad. I was just covered. I let out a holler to the navigator, and he said, "What's wrong, skipper?" I said, "Who the hell turned that spout into wind?"

Airman: As your number of trips went up and you hit the magic number, you quit telling people how many trips you had in for your tour and told them how many you had left. You used to say you had six ops in or seven ops in. But when you started coming close you'd say, "Only ten to go," and "five to go," and "three to go." It was a different thing. We got very serious.

Air Gunner: I had a very short career. I had my nineteenth birthday on December seventh, 1943, but I was shot down on December fourth, 1943.

Air Gunner: We were stationed in Normandy, and an Me 109 came across scouting us, but he had his undercarriage down and he was smoking, so I didn't know what the hell to think. But I picked up my rifle and fired. I had some tracer ammunition, and, goddammit, some of those tracers came close. I'm not sure if he was going to land; he was coming in, but his undercarriage might have been shot down. Then he kept on going and climbed, and a few moments later he bailed out and the aircraft spun in. He *had* been smoking, so maybe somebody else had shot him. But I did fire three shots, for our side.

Pilot: We used to go drinking, lots of drinking. My God, when I think of that. It was a case of relieving the tension I suppose. You were under an awful lot of tension, but at that age and in that physical condition it's a lot easier to bear, and you didn't think about it. You just didn't think, and I guess that's one reason that *young* men fight wars.

You remember the good times; you don't remember the times you were scared. And you have to be scared; if you're not scared, then you're stupid. When you're scared, then your senses are far more alert and you are a lot more careful. And there was an element of ignorance involved with a lot of us. Here I was, at the age of nineteen when I picked up my crew, responsible for the lives of six other people. But that thought never entered my head, because this was just a job.

*****

Official, Finance Department: When the war was over and we were settling our accounts, the British owed us about $450,000,000, but the British hadn't been keeping careful track of their financial commitments. After lend-lease had started, they were getting some planes under that and putting them into the Plan. So towards the beginning, in the first two and a half years of the Plan when the British contribution was seriously deficient, it had been substituted by planes from the United States. Because they weren't able to produce enough planes, the British ended up owing us a couple of hundred million dollars. For the second part of the Plan they accepted the liability for a share of the total expense, and it was understood that they would meet their share by supplying more aircraft. But by then the cost of building training aircraft had been substantially reduced, and thus their value, and Britain ended up owing us quite a bit more in cash to compensate.

When it finally came down to the wire, we said, "You can just forget it," and we wrote off the whole $450,000,000. But we never made any move towards that until early in '46 when we toted up their whole bill, and we threw it in then as part of their general settlement for the war.

In the end, the cost of the Plan to Canada was about one and two-thirds billion dollars.

Ground Staff, RCAF: *I was sent down to Swift Current with 100 airmen to start storing aircraft. That became a storage unit, and they were sending them in from all over the country. We wound up with 1,100 and some odd at that one field. The hangars were full so we stored most of them outside. We dug holes in the dirt and dropped the wheels in, and put screw pickets down for the tailplanes. Then we covered the engines, dropped all the controls, locked the cabins and coupe tops, and that was it. Eventually they were handed over to War Assets.*

Airman, RCAF: It's incredible that the Air Training Plan got under way so quickly and so effectively. Canada had an extraordinary number of people trained in an extremely short time. How they got so many together to operate the thing reflects a great deal of credit on the people that did the initial organizing. I personally walked out of civvy street and in less than a year was flying on operational aircraft. That's quite an achievement, and certainly says a lot for the calibre of Canadian ability.

Engineering Officer, RCAF: C.D. Howe was the key man in the co-ordination of production. He was the Minister of Munitions and Supply, a keen businessman, a hard driver, but inspirational in his ways. He was the one who mobilized industry to really pitch in. As Lord Beaverbrook was to production in England, Howe was to Canada. He had the ability to understand if the individual with whom he was dealing was on the ball, that he was on the job and that he was achieving. He would then say, "That's it!"... and once he said that, you knew that he was behind you and that he would cut all possible corners to get the job at hand completed.

As Minister of Defence, Ralston determined that the Department of Defence as such would not be able to accomplish what had to be achieved in the form in which the Department then existed. Mackenzie King agreed, and created a Minister for Air, a Minister for the Navy, and a Minister for the Army as a wartime measure. And so Chubby Power was appointed the Minister for Air.

Airmen, RAF: For the Royal Air Force the beginning of the pipeline was in Britain in twenty or thirty different grading schools, where they did the elementary try-out flying. The recruits all had to be collected and assessed, and then somebody had to work out the logistics of a troopship for them. They had to come to Canada. There had to be a post: "Where are they to go? What's the time program at this EFTS? When's that course going to SFTS? Where does everybody fit in?" You couldn't have 10,000 people sitting and cooling their heels in one spot; they had to be kept on the move.

Airman, RCAF: You had deadlines at all the EFTS and all the SFTS. A course came in on a date, and it had to go out on a date, because there was another batch coming in. Where are *they* coming from? Somebody's going to take *their* beds. It was all planned like that. When we finished a course we got a week's leave, then there was a vacancy for us at two or three SFTSs because the people there had been moved along. It was a great game of turning the wheel, and it was extremely well organized.

Airman, RCAF: Looking at figures of graduates of pilots and navigators and the whole complex, it's overwhelming. I couldn't imagine that that many people had graduated. But when you look at your own little course of thirty people going through in six weeks, and another thirty behind you in six weeks, it would all build up. I never looked at it as a British Commonwealth Air Training Plan; I always thought of it as a lot of elementary schools, a bunch of AOSs, and some service schools. Everywhere you went there was a station, but I don't think you ever sat down and said, "Holy Christmas, there's a lot of people going through this thing."

215

Airman, RCAF: *All I knew was that at every stage of the way there were people who were trained and equipped to teach me skills that I didn't have and which I would need. And with successive accomplishments I went from stage one to stage two, to stage three, and so on. There wasn't any real, deep appreciation of what a fantastic organization had made me from nothing into a pilot, and "I can now fly an aircraft against the enemy." That sort of thing was never in my mind.*

Airman, RCAF: *We weren't really an industrial nation and we weren't in the forefront where the aviation world was concerned. Canada's aeroplane world to that time was daredevil bush pilots in the north. Then suddenly, with the stroke of a pen and some very brilliant people, they made this thing work. When you look at what they were able to turn out on such short notice, you realize the magnitude of the achievement. Somebody signed a piece of paper and, damn it, they made it work.*

Station CO, RCAF: *It was just marvellously well organized, a great credit to Canada. I don't think it could have been done anywhere else but in this country. We had Australians and New Zealanders and we had Englishmen—a tremendous mixture of people from all over the Commonwealth. The maintenance was excellent. We were doing over 14,000 hours a month on my station. I don't know how the hell it ever got organized from Ottawa, but it's always been a marvel to me how well it worked.*

CANADA HAD BROAD HORIZONS BUT LITTLE ELSE

PR Officer, RCAF: That Plan was a thrilling thing that happened, an amazing thing. If it hadn't happened, then you ask yourself, "Well, what?" If New Zealand and Australia and Canada had sent their pilots over to England after they had trained a few, as soon as they possibly could, and they had gone over to France and been outnumbered, where would we have been? It's highly possible we wouldn't have won that war.

Because of its nature the British Commonwealth Air Training Plan is unique in any military form. There never had been, in any one country, quite that concentrated a gathering of people to train, and be trained, and to create the whole machine. It was an entirely new thing, anywhere for anything, military or commercial. Canada had the broad horizon of air and land, but little else. It had never done anything of this scope, to have to build the plant and train the men to fly, who would then train thousands of other men to fly.

The British had the guts to decide, "We can hold out, just for two years. But, by God, at the end of two years we're not going to have a plane or a crew left, and then we'll lose the bloody war." The greatest cost of the BCATP, which probably kept us from losing the war, was paid by the English, because they were the ones that were being whacked by those raids while the Plan tooled up for maximum output. But the Brits knew what they were doing, and they didn't complain.

Airman, RCAF: I had no concept of size at all. All I knew was that there was a Plan and there was a place for me. I wanted to be the elite; I wanted to be in the Air Force, because they were taking the best boys, the best minds, the best bodies, and I wanted to be there.

Airman, RCAF: While one must be tremendously amazed at the large number of aircrew that went out, the real heroes were the ground staff that made it all work. Every time one of those Moths or Ansons or Harvards or Bolingbrokes got off the ground, somebody had serviced it, somebody had been a top-notch mechanic. There were weather people; there were aerodrome control people; there were endless numbers of people behind the actual flying training operations; where in the hell did they all come from? That scheme was a marvellous piece of machinery.

Ground Staff, RCAF: You knew that you were part of a hell of a big organization that was there to get people knowledgeable on aeroplanes and on procedures regarding aeroplanes to assist in the big war. You were only a wee peg in the whole damn set-up, but you kinda felt, "I guess I'm doing a job here, too, that's helping to do the thing up right to get on with this war. It's a big one; keep it moving."

Airman, RCAF: It's only in later years that you begin to wonder, "Who stuck all those aerodromes around? Where did the training staff come from? Where did the aeroplanes come from? And where did all the students come from, and the cooks, and the bands, and the clerks?" Good Lord, it was really an outstanding achievement. We had nothing to do with forming it, we were merely the nuts and the bolts that went through the grinders and came out the other end, and when the big show was over, most of us were in such a damn hurry to get out and get on with our private lives that we really didn't pay too much attention to it.

Airman, RCAF: Trainees were brought from literally around the world and put through the programs. It was an incredible job, an ingenious concept, to decide that Canada was big enough and had all the requirements to establish and maintain a massive training program to arm the Allies with airmen.

Officer, RAF: With all the air raids and the blackouts, there was no way they could have trained that volume in England. They could never have got the quantities they wanted. I did some instructing in England, and we'd have the odd flare path out, but if an air raid started they'd have to pack everything up. The Plan was a brilliant idea.

Instructor, RCAF: We had a course of RAF come over who had trained on Tiger Moths in England, and on their last day of flying they'd been shot up by a Ju 88. They lost a couple of students and some of the others got pretty badly shaken up, so we had the jitteriest bunch of fellows you ever saw. Here, they had a safe training area.

Officer, RAF: In those days everything had to be done by ship and train and truck. And yet, they transported these pupils over vast distances, and people got to their training stations, got their courses, and then were moved on, and they never arrived when there weren't meals and beds available for them. People came from all sorts of different places at unpredictable times. Ships were rerouted in mid-Atlantic or had to make detours, but the trains were there to meet them. There was some fantastic organization involved in the job of co-ordination. That kind of logistics just boggles the mind.

Airman, RCAF: Given the experience level and the diverse backgrounds of so many people, somebody must have had their act together to produce aircrew of the calibre that were turned out. I take my hat off to the instructors; they must have been well selected and they must have been pretty damn competent, and the system must have been pretty good. When you think about what you learned in the space of less than a year, the system itself must have been well thought out, because it was pretty fantastic.

Station CO, RCAF: It was a tremendous success, because everyone that was involved in the machine worked hard and was well motivated. The progress or lack of progress of the war was the big factor in the way the Allies felt: "We have to win this goddam thing!" It was ingrained in them, or that would be the end of our way of life.

Navigator, RCAF: At 24 years of age I thought their whole system stunk; they didn't know a good man when they saw one! At that one stage when I was washed out I thought it was awful, but when I was going through they had the choice of many, many young men for pilots, so why wouldn't they take the best, the ones that trained better and didn't do stupid things? If I had got pilot Wings, I never would have got out of Canada. I'd have ended up a stooge pilot somewhere, and that wouldn't have made me very happy.

I'm 65 years old now, and I look at this thing a hell of a lot different than I did when I was 24 years old. I think it was fantastic.

Officer, RAF: The British Commonwealth Air Training Plan made a very special contribution to the Allied cause, that can only be appreciated by those who were aware at the time of what was happening.

Airman, RCAF: It was probably one of the most brilliant manoeuvres of the war.

EPILOGUE

In the summer of 1982 I joined two of my very good friends who had been through the Training Plan with me, and we drove down to Weyburn, down to Estevan, down to Assiniboia, up to Moose Jaw, to Regina, to Carberry, to Portage la Prairie, all the places we'd been in and out of during the war. Some of the stations were easily recognizable, Weyburn was almost all there, and at Estevan there were just bits and pieces, but we were able to walk around and everyone had stories about incidents or the stations.

We went to graveyards. At Estevan, with all those pupils flying over a period I suppose of three and a half years, the only fatal accident they had was when an instructor had taken another instructor up for a test flight and persuaded three of the groundcrew to go with them. These guys wanted to go, because they got a dollar or something extra for flying pay. Well, like many of these young instructors they got showing off, and they flew that Anson into the ground, and wrote them all off.

So there they are, in that little patch of grass up in Estevan, on a hill where it says "RAF" on the cemetery. The one's in the middle with guys on either side. You read the thing, nineteen, eighteen and a half, and an inscription that their mother had sent out, "Dear Son, sadly missed." You know, it gives you a choke when you read a thing like that. And the one at Assiniboia—there are nineteen of them there—all but one were dual crashes. I hadn't realized that they had inscriptions on them. When you go to the funeral there's no stone, and I suppose that's why. But they all had little inscriptions that their families had sent out. And I hadn't realized they were all as young as that. You wonder where they'd be today, forty years later. But there they are, snuffed out. Not their fault—just a stupid instructor overestimating his skill and the ability of the aircraft. Just wrote these guys off . . . terrible.

There were several hundred killed in Canada in those war years—the years of the Plan—and they're scattered from one end of the country to the other. I've been to places where you find large central memorials which not only list the names, but also commemorate the enterprise in which these men died. But there's not a place like that for the British Commonwealth Air Training Plan.

I got my Wings; I got my gongs; I got my sort of bellyful of war after I had done my training, as many of the other fellows did. But these *other* lads got nothing except death. And it really doesn't matter whether you're pranging in a Tiger Moth over Neepawa or in a Halifax over Berlin—you're just as dead. And these young fellows were all just as loyal, just as courageous, just as young, and just as prepared to go and do those big and glorious deeds as those that got to do them, but they never got the chance and they got none of the rewards.

They guy that was killed later usually had a few weeks, or months, or years of glory beforehand—whether that made it worthwhile or not is another matter—but these kids got nothing. They got no Wings; they got no recognition; they got no memorial, they got no gongs. Nobody ever talks about them any more . . . and I think they should.

INDEX

Airspeed Oxford, 88, 175
Armstrong-Whitworth Siskin, 4, 5, 19, 23
Assiniboia, Saskatchewan, 70, 202, 219
Assiniboine Park, Winnipeg, 148
Avro Anson, 42, 46, 48, 54, 86, 88, 89, 91-94, 96, 97, 103, 106, 110-113, 116, 117, 120, 126, 127, 131, 134, 137, 140, 143, 145, 162, 166, 175-177, 183, 196-199, 213
Aylmer, Quebec, 38

Bell, Ralph, 34
Bennett, R.B., 18, 19
Berlin, Germany, 219
Bishop, Billy, 55, 164, 166, 167
Black, Conrad, 34
Boeing School of Aeronautics, California, 58
Boundary Bay, British Columbia, 169
Bournemouth, England, 210, 212
Bowden, Alberta, 202
Brampton, Ontario, 171
Brandon, Manitoba, 50
Brantford, Ontario, 172
Bristol Beaufighter, 211
Bristol Blenheim, 120
Bristol Bolingbroke, 120, 176, 198

Cagney, James, 166, 167
Calgary, Alberta, 24, 36, 50, 73, 90, 97, 170
Camp Borden, Ontario, 18, 19, 21, 23, 24, 35, 36, 44, 153, 168
Canadian Pacific Airlines, 112, 118
Canadian Pacific Railway, 139
Cap de la Madeleine, Quebec, 74
Captains of the Clouds, 166, 167
Carberry, Manitoba, 219
Carmichael, Harry, 34
Cartierville, Quebec, 39
Caughnawaga Indian Reserve, Quebec, 126
Cessna Crane, 73, 88, 97, 193
Chamberlain, Neville, 24
Chatham, New Brunswick, 116
Chicago, Illinois, 144, 171
Clinton, Ontario, 171

Collard, Dick, 29
Curtiss Mohawk, 44
Curtis, Wilf, 29

Dafoe, Saskatchewan, 173
Dartmouth, Nova Scotia, 53
Dauphin, Manitoba, 30, 93
Davis, R.C., 4, 5
Debert, Nova Scotia, 54, 55
De Havilland Moth, 22, 36, 37, 47, 49, 50, 60-62, 66, 71-75, 101, 103, 120, 132, 164, 180, 185, 189, 195, 201-203, 213, 217-219
Detroit, Michigan, 171
Dexter, Grant, 34
Dorchester, Nova Scotia, 159
Duke of Windsor, 97, 98
Duncan, James, 34
Dunkirk, France, 138
Dunnville, Ontario, 38, 168

Edmonton, Alberta, 36, 37, 52, 112, 175, 177
Estevan, Saskatchewan, 32, 218, 219

Fairey Battle, 39, 42, 47, 85, 111, 120, 125, 164, 175, 213
Fleet Finch, 37, 39, 47, 59, 61, 73, 90, 93, 101, 111, 205
Fleet Fort, 164, 189, 193
Ford Trimotor, 4, 5, 19
Fort Erie, Ontario, 39
Fort Peck Dam, Montana, 198
Fort Simpson, North West Territories, 4
Fort William, Ontario, 201
Fullerton, Elmer, 4

Gable, Clark, 153
Golden, British Columbia, 156
Gone With The Wind, 153
Greenock, Scotland, 210
Gulf of St. Lawrence, Quebec, 112, 117

Hagersville, Ontario, 98, 153
Halifax, Nova Scotia, 29, 32, 139, 206-209
Hamilton, Ontario, 39

Handley Page Hampden, 39, 111
Hawker Hurricane, 4, 5, 188, 211

James, Harry, 173
Junkers F-13, 4

Kaufman County Airport, Texas, 145
King, Mackenzie, 18, 21, 22, 24-27, 33, 215
Kingston, Ontario, 193, 213

Leckie, Robert, 35
Lennoxville, Quebec, 156
Lethbridge, Alberta, 62, 121, 126, 176
Little Norway, Toronto, 44, 150-152
Lockheed Lodestar, 54, 144
London, Ontario, 126, 171

Malton, Ontario, 39, 69, 74, 89, 104, 111, 127
Medicine Hat, Alberta, 48, 89, 202
Moncton, New Brunswick, 30, 141-145
Montreal, Quebec, 17, 29, 54, 59, 107, 119, 141, 165, 169, 199, 206, 207
Montreal Light Aeroplane Club, 59
Moose Jaw, Saskatchewan, 219
Mossbank, Saskatchewan, 30, 121
Muskoka, Ontario, 133, 135, 151, 153

Neepawa, Manitoba, 60, 219
Newcastle, New Brunswick, 116
New York, New York, 208
North American Harvard, 17, 38, 42, 48, 88-93, 97, 110, 143, 155, 166, 175, 183, 188-190, 192, 196, 200, 205, 213, 217
North American Yale, 38, 42, 88, 110
Northern Saskatchewan Flying Club, 61

Orangeville, Ontario, 193
Ottawa, Ontario, 30, 38, 45, 112, 121, 166, 169, 187, 216
Ottawa Flying Club, 60
Owen Sound, Ontario, 196

Pearl Harbor, Hawaii, 144, 168

Plumas, Manitoba, 134
Portage la Prairie, Manitoba, 32, 77
Power, C.G., 26, 215
Prince Albert, Saskatchewan, 61, 69, 132, 195
Pugwash, Nova Scotia, 159

Quebec Airways, 113
Quebec City, Quebec, 107, 108

Ralston, J.L., 26
Regina, Saskatchewan, 23, 36, 93, 141, 219
Roberts, Lofty, 19
Rockcliffe, Ontario, 53, 165
Ryder, Lady, 212

Saint John, New Brunswick, 29
St. Thomas, Ontario, 28, 29, 38, 121, 152
Saskatoon, Saskatchewan, 29, 39, 50
Stearman PT-27, 72, 101
Supermarine Spitfire, 73, 76, 188, 211
Swift Current, Saskatchewan, 143

Toronto, Ontario, 17, 28, 39, 50-52, 55, 56, 69, 104, 112, 118, 127, 153, 172, 183, 207
Trans-Canada Air Lines, 17, 69, 74, 76, 92
Trenton, Ontario, 19, 21, 22, 35, 36, 77, 97, 100, 101, 168, 189, 205
Truro, Nova Scotia, 139

Uplands Airport, Ottawa, 17, 74, 89, 199, 200

Vancouver, British Columbia, 4, 21, 103, 146, 147, 152
Victoriaville, Quebec, 55, 173, 174

Westland Lysander, 120, 124
Weyburn, Saskatchewan, 219
Windsor, Ontario, 207
Winnipeg, Manitoba, 19, 22, 32, 34, 93, 98, 111, 119, 141, 149, 161, 172
Wolseley, Saskatchewan, 22

Yorkton, Saskatchewan, 30

ACKNOWLEDGMENTS

The author expresses his appreciation to the following for their unselfish generosity in sharing their time, their photographs, and their memories, which have contributed to the completion of *The Plan*.

Aass, H.	Douglas, A.	Mears, T.C.
Anderson, C.W.	Douglas, R.	Melnick, J.
Angus, D.	Edmonton Flying Club	Merrilees, T.L.
Antenbring, A.	Fether, E.	Moncrieff, E.H.
Armstrong, C.	Finland, M.	Moore, E.
Aylard, D.	Finley, H.	Moore, G.
Bennett, W.J.	Fry, M.	Newson, A.
Bladon, R.	Graham, E.G.	Nicolaiff, J.
Blaine, D.S.	Grimes, T.	Nyhuus, J.
Bova, G.	Hansen, A.	Ottawa Evening Citizen
Bradford, R.W.	Hantiuk, J.D.	Pennie, A.
Bridgemen, P.	Hatch, F.	Pickersgill, J.W.
Briggs, E.J.	Hawkes, D.	Powell, B.
Brown, J.	Irwin, C.	Public Archives of Canada
Brown, M.	James, B.	RCAF Association, 700 Wing
Browne, K.	Janaga, M.	Richardson, H.
Bruchet, I.	Jennings, J.	Ricketts, H.
Bryce, R.B.	Johns, C.F.	Robinson, W.
Butler, J.	Johnstone, J.	Shortt, N.
Calder, J.	Kellow, B.	Skistod, T.
Callingham, J.	Kennedy, A.J.	Sambrooke, D.
Cameron, E.	Kennedy, E.	Smith, I.N.
Campbell, H.	Keir, J.	Stewart, W.
Campbell, J.K.	Klimpke, W.	St. John, B.
Carlson, W.	Langtry, J.	Taylor, M.
Carr, J.	Lavigne, J.	Templeton, J.I.
Cheffins, B.	Lennon, L.	Tibbs, S.J.
Clay, D.	Lovell, J.A.B.	Todd, A.
Cockle, A.	Lyon, J.H.	Towne, E.
Congdon, B.	MacCaulay, E.H.I.	Vandem, B.
Conquist, L.	MacDonald Bros.	Vincent, A.
Coyne, A.	MacKenzie, K.	Walker, A.
Curtis, M.	MacWilliam, E.	Walker, J.
Dalzell, O.	Machan, F.	Wanless, W.
Dancose, P.	Mallandaine, J.	Wartime Pilots & Observers Assoc.
Davis, F.	Martin, A.	Warren, O.
Dept. of National Defence	Martin, H.	Watts, W.
Dickey, B.	Martland, J.	Webb, G.
Dickey, W.	McAllister, T.	Wilson, B.
Dodds, E.	McDiarmid, A.	Wilson, D.
Donovan, J.	McMillan, S.	Wishart, H.
		Wood-Sannon, D.

OLD FRIENDS

The following photographs were supplied
by the Public Archives of Canada:

Page	Photo Number
20	PA062735
20	PA062259
21	PA062837
23	PA062441
25	PA062297
28	PA063635
53	PA136880
59	PA062534
133	PA115427
151	PA136881